AI Projects in PyTorch

Hands-On Projects in Vision, Text, and Generative Models

Siddhesh Prashant Chaubal

Apress®

AI Projects in PyTorch: Hands-On Projects in Vision, Text, and Generative Models

Siddhesh Prashant Chaubal
Thane, Maharashtra, India

ISBN-13 (pbk): 979-8-8688-2116-5   ISBN-13 (electronic): 979-8-8688-2117-2
https://doi.org/10.1007/979-8-8688-2117-2

Copyright © 2025 by Siddhesh Prashant Chaubal

This work is subject to copyright. All rights are reserved by the Publisher, whether the whole or part of the material is concerned, specifically the rights of translation, reprinting, reuse of illustrations, recitation, broadcasting, reproduction on microfilms or in any other physical way, and transmission or information storage and retrieval, electronic adaptation, computer software, or by similar or dissimilar methodology now known or hereafter developed.

Trademarked names, logos, and images may appear in this book. Rather than use a trademark symbol with every occurrence of a trademarked name, logo, or image we use the names, logos, and images only in an editorial fashion and to the benefit of the trademark owner, with no intention of infringement of the trademark.

The use in this publication of trade names, trademarks, service marks, and similar terms, even if they are not identified as such, is not to be taken as an expression of opinion as to whether or not they are subject to proprietary rights.

While the advice and information in this book are believed to be true and accurate at the date of publication, neither the authors nor the editors nor the publisher can accept any legal responsibility for any errors or omissions that may be made. The publisher makes no warranty, express or implied, with respect to the material contained herein.

    Managing Director, Apress Media LLC: Welmoed Spahr
    Acquisitions Editor: Celestin Suresh John
    Coordinating Editor: Gryffin Winkler

Cover designed by eStudioCalamar

Cover image by freepik (freepik.com)

Distributed to the book trade worldwide by Springer Science+Business Media New York, 1 New York Plaza, New York, NY 10004. Phone 1-800-SPRINGER, fax (201) 348-4505, e-mail orders-ny@springer-sbm.com, or visit www.springeronline.com. Apress Media, LLC is a Delaware LLC and the sole member (owner) is Springer Science + Business Media Finance Inc (SSBM Finance Inc). SSBM Finance Inc is a **Delaware** corporation.

For information on translations, please e-mail booktranslations@springernature.com; for reprint, paperback, or audio rights, please e-mail bookpermissions@springernature.com.

Apress titles may be purchased in bulk for academic, corporate, or promotional use. eBook versions and licenses are also available for most titles. For more information, reference our Print and eBook Bulk Sales web page at http://www.apress.com/bulk-sales.

Any source code or other supplementary material referenced by the author in this book is available to readers on GitHub (https://github.com/Apress). For more detailed information, please visit https://www.apress.com/gp/services/source-code.

If disposing of this product, please recycle the paper

Dedicated to my amazing wife, Sayali.

Table of Contents

About the Author .... **xiii**

About the Technical Reviewer .... **xv**

Acknowledgments .... **xvii**

Introduction .... **xix**

Chapter 1: Introduction to Machine Learning **1**

   AI and ML 1

   Machine Learning with an Example 2

      A House Price Prediction Problem 2

      Linear Regression 2

      Multivariate Linear Regression 4

      Gradient Descent 5

      Feed-Forward Neural Networks 5

   Machine Learning Primer 8

      Train-Test-Validation Split 8

      Model Evaluation 10

      Domains of Machine Learning 11

   Data Collection 12

   Data Preprocessing 12

   Feature Engineering 13

   Model Training 15

      Training Neural Networks Using Gradient Descent and Backpropagation 15

      Updating Parameters 16

      Gradient Computation 17

Table of Contents

Overfitting and Underfitting 19

Summary 21

Chapter 2: Tensors in PyTorch 23

Introduction to Tensors 23

    Slicing 27

Initializing Tensors 31

    From Another PyTorch Tensor 31

    With Constant Values 31

    From a NumPy ndarray 32

Computational Graphs and Gradients 36

    Gradient Descent in Neural Networks 38

Exercises in Tensor Manipulations 40

Summary 48

Chapter 3: Image Classification Using Convolutional Neural Networks 51

What Is Image Classification 52

    History 52

Convolutional Neural Networks (CNNs) 53

    Convolutions 53

    Vertical and Horizontal Edge Detection 54

    Activation Functions 55

    Strides 56

    Padding 56

    Dimension Calculations 56

    Dropout Layer 60

The Image Classification Project 61

    Dataset 61

    Main Components of the Project 61

    DataLoaders 63

    Data Exploration and Visualization 64

    Model Definition 66

Model Training Loop	68
Putting It All Together	70
Refining the Model	71
A More Challenging Dataset	81
Data Loading and Augmentation	81
Transfer Learning: Fine-Tuning a Pretrained Model	87
Summary	90

Chapter 4: Introduction to Natural Language Processing: Building a Text Classifier 93

Natural Language Processing (NLP)	94
What Is NLP	94
Popular Problems Solved by NLP	94
Preprocessing	95
One-Hot Encodings and Embeddings	96
Text Classification	97
A Typical NLP Modeling Approach	97
The Encoder Architecture	98
RNNs and LSTMs (Optional)	99
The Transformer Architecture	101
Sequence-to-Sequence Models	104
The Text Classification Project	105
Imports	106
Dataset	106
Dataset Statistics	107
Preprocessing	110
Data Loader	113
First Model: Training the Embedding Layer	115
Training, Validation, and Testing	118
Putting It All Together	121
Results	122

TABLE OF CONTENTS

    Second Model: Using Pretrained Word Embeddings Using Word2Vec 122

    Alternative: Training the Word2Vec Model from Scratch 126

  Text Classification with Pretrained Hugging Face Transformers............................ 127

    BERT (Bidirectional Encoder Representations from Transformers) 128

  Summary.. 132

Chapter 5: Practical Natural Language Processing with Hugging Face 135

  Hugging Face 136

    What Is Hugging Face.. 136

    The Model Hub 137

    The Dataset Hub 137

    The Pipeline API: Instant Inference 138

    Fine-Tuning: Specializing Pretrained Models 138

  Text Classification 139

    Project 1: Zero-Shot Classification 139

    Project 2: Fine-Tuning Pretrained Models for Emotion Classification............... 141

  Summarization.. 153

    Extractive Summarization.. 153

    Abstractive Summarization 154

    Summarization Project 1: Summarizing a Wikipedia Article............................ 154

    Summarization Project 2: Conversational Chat Summarizer 158

  Question Answering 173

  Summary.. 175

Chapter 6: Building a Language Model for Storytelling.. 177

  Introduction to Language Modeling 177

    N-Gram Models 178

    Deep Learning Models: RNNs and LSTMs 178

    The Magic of Transformers... 179

  Transformer Decoder-Only Architecture... 179

    Self-attention Mechanism 179

    Multi-head Attention... 182

    Positional Encodings 182

    The Decoder Block 183

    The Final Transformer Language Model 183

  Project: A Story Writing Language Model from Scratch 185

    A Note on Computing Resources 185

    Imports 185

    Dataset 186

    Train-Test-Validation Split 188

    Tokenization 188

    Batching 190

    A Simple Baseline: Bigram Model 192

    Decoder-Only Transformer Model 202

    Decoder-Only Model with Word-Level Tokenization 212

    Further Improvements 214

    Troubleshooting 215

  Summary 215

Chapter 7: Audio Classification with PyTorch 219

  Understanding Sound 219

    How Computers "Hear" 220

    Classifying Digital Waves 221

    Mel Spectrogram: Turning Sounds into Model-Friendly Images (Optional) 222

  Project 1: Speech Command Recognition 224

    Imports 225

    Dataset 225

    Data Processing Parameters 227

    EDA 227

    Data Preprocessing 231

    Model Architecture 234

    Model Training and Evaluation 237

TABLE OF CONTENTS

Project 2: Speech Emotion Recognition (SER) 241
    Imports 242
    Dataset 244
    Further Improvements 271
Summary 271

Chapter 8: Recommender Systems with PyTorch 273
What Are Recommender Systems 273
Collaborative Filtering 274
Recommender Systems Project 276
    Imports 276
    Dataset 277
    Data Analysis and Visualization 279
    Approach 1: User–User Collaborative Filtering 282
    Approach 2: Matrix Factorization 290
Summary 295

Chapter 9: Image Captioning with PyTorch 297
Multi-modal Models 297
    The Encoder-Decoder Architecture 298
Image Captioning Project 299
    Imports 299
    Downloading the Data 300
    Train-Test-Validation Split 302
    Reading Captions 303
    Exploratory Data Analysis (EDA) 305
    Vocabulary and Tokenization 310
    Dataset Class 312
    Image Transforms 313
    DataLoaders 314
    Model Architecture 317
    Model Training 326

TABLE OF CONTENTS

    Model Inference: Generating Captions for Images 328

    Model Evaluation 331

  Further Improvements 334

  Summary.. 335

Index.. **337**

About the Author

 **Dr. Siddhesh Prashant Chaubal** has dedicated his career to building and studying intelligent systems — from cutting-edge research in artificial intelligence to large-scale machine learning platforms powering real-world applications. Currently, he works as a Research Scientist at Dream11 in Mumbai. In earlier roles, he has served as a Staff Engineer at Qualcomm India and an Applied Scientist at Amazon in Seattle. He holds a B.Tech. in Computer Science from IIT Bombay and a PhD from the University of Texas at Austin, where his research explored theoretical aspects of computer science and machine learning. His work has been published in leading international conferences such as CIKM and MFCS. When not immersed in AI, he enjoys reading, playing chess, or listening to music.

About the Technical Reviewer

Shibsankar is currently working as a Senior Data Scientist at Microsoft. He has 10+ years of experience working in IT, where he has led several data science initiatives, and in 2019, he was recognized as one of the top 40 data scientists in India. His core strength is in GenAI, deep learning, NLP, and graph neural networks. Currently, he is focusing on his research on AI agents and knowledge graphs. He has experience working in the domains of foundational research, FinTech, and ecommerce.

Before Microsoft, he worked at Optum, Walmart, Envestnet, Microsoft Research, and Capgemini. He pursued a master's from the Indian Institute of Technology, Bangalore.

Acknowledgments

I am grateful to Santanu Pattanayak, my manager at Qualcomm, who first introduced me to Apress. His passion for learning and writing has been a true inspiration and a significant influence on my own journey.

My sincere thanks also go to the entire editing and production team at Apress for their invaluable guidance and support. In particular, I am grateful to Celestin, Shibsankar, Nirmal, and the rest of the team, whose efforts have been instrumental in shaping this book into its final form.

I am profoundly thankful to my parents, who have been a source of strength and encouragement throughout my life.

Above all, I would like to express my deepest gratitude to my wife, whose encouragement, support, and patience have been indispensable to the completion of this book. She has also been kind enough to prepare many of the illustrations of this book, which add clarity and support to the explanations.

Introduction

This book is primarily meant as a segue into artificial intelligence for software engineers with hands-on projects. It also serves as a guide to mastering PyTorch, which is one of the most popular frameworks for deep learning. The initial chapters cover the fundamentals of machine learning and PyTorch. Subsequently, it goes into different domains of AI, namely, computer vision, natural language processing, audio classification, and recommender systems. Each domain is brought to life through one or more end-to-end projects.

Who Will Benefit from This Book

This book is primarily meant as an introduction to AI and its different domains for readers familiar with Python. As such, anyone with an intermediate grasp of Python who is interested in venturing into AI will benefit from reading this book. Specifically, software engineers, or other professionals who dabble in Python for their work, are one of the primary audiences. Curious students and enthusiasts can also benefit, discovering practical ways to begin their journey in AI.

How This Book Is Organized

The book starts with two background chapters that build the foundations for all the projects to follow. Chapter 1 gives a primer on AI and machine learning, assuming no prior knowledge. Chapter 2 dives into the nitty-gritties of PyTorch, with several programming illustrations and exercises at the end of the chapter. The subsequent chapters each contain one or more hands-on projects, typically beginning with a section that explains the necessary background of the domain before moving into the project.

INTRODUCTION

Computer Vision

Chapter 3 introduces the field of computer vision with a project on image classification using convolutional neural networks (CNNs).

Natural Language Processing

Chapters 4-6 focus on natural language processing (NLP). Chapter 4 begins with an introduction to NLP along with common preprocessing steps such as tokenization, numericalization, padding, and truncation. It also explains the evolution of different modeling approaches, from RNNs to transformers, before moving into a text classification project using various strategies. Chapter 5 introduces modern NLP, teaching various aspects of the Hugging Face ecosystem, and tackling four different NLP tasks with pretrained models, including fine-tuning. Chapter 6 builds a transformer-based language model for storytelling.

Audio Classification

Chapter 7 introduces the audio processing domain, guiding the reader with a project in audio classification.

Recommender Systems

Chapter 8 covers the foundations of recommender systems and includes a hands-on project.

Multimodal Models

Chapter 9 walks the reader through an image captioning project using a multimodal model that combines vision and NLP.

How to Read This Book

This book is meant as a practical guide, so I strongly recommend running all the code in each chapter step by step. Treat the chapters like a lab notebook: experiment with different parameter settings, try variations, comment out parts of the code to see what breaks, and follow your curiosity. The more actively you explore, the more deeply you will master these ideas.

The first two chapters are foundational and warrant careful study – if you are new to PyTorch, be sure to complete all the exercises at the end of Chapter 2. If you are already comfortable with ML or PyTorch, you may skim them. The remaining chapters can generally be read in any order, though I recommend starting with Chapter 3 before going further, as it introduces additional practical PyTorch and ML concepts.

CHAPTER 1

Introduction to Machine Learning

Machine learning is a rapidly expanding field – both in academic research, with thousands of papers published every month, and in the development of revolutionary tech products like ChatGPT and Veo. This chapter explains the basics of machine learning, enabling you to understand the fundamentals behind these advanced technologies, some of which we will discuss in the later chapters.

Chapter 1 begins with a brief introduction to artificial intelligence (AI) and machine learning (ML). Next, we take up a simple example to build intuition for the basics of machine learning, where we introduce linear regression and neural network algorithms. We then explain some of the ML concepts in depth, including practical subtleties like data collection, preprocessing, feature engineering, etc. We then move on to model training, explaining the mechanics of the gradient descent algorithm in detail. Finally, we conclude this chapter by explaining the ideas of model overfitting and underfitting.

AI and ML

Artificial intelligence (AI) has been a major topic of discussion across both social media and scientific forums over the past decade. AI encompasses all endeavors toward emulating human behaviors in machines, whether in self-driving cars, factory robots, or AI chatbots like ChatGPT. For developers, AI often takes the form of software that demonstrates humanlike cognitive abilities.

The two main domains of AI where major strides have been made recently are computer vision and natural language processing. On a high level, the former seeks to emulate the functionalities of the human eye, while the latter develops a computer's comprehension of human languages. Most of the projects in this book will belong to one of these two domains of AI (or both). We will be working with data-driven algorithmic approaches to these problems, commonly referred to as machine learning (ML) in literature, and henceforth, we will use these two terms (AI and ML) interchangeably.

Machine Learning with an Example
A House Price Prediction Problem

Let us say that a one-bedroom house in a certain neighborhood sells for 100K, a two-bedroom house sells for 200K, and a three-bedroom house sells for 300K. Now, as an ML expert, you are asked to predict the price of a four-bedroom house in that same locality (with no additional info). What would you say? You would not need to write a Python program to predict the price of 400K for this house. In fact, a fourth grader could answer this question without batting an eyelid. But what methodology does our brain really use to come up with this answer? This question is worth asking because only then would we still be able to predict the price if the numbers were less obvious.

Let us make it less obvious: say the price of a one-bedroom was 120K, and that of a two-bedroom was 220K, and that of a three-bedroom was 320K. A little more thought will tell you that a similar method works after subtracting a constant value of 20K from each price. So, you will still be able to correctly predict 420K without using a pencil.

Linear Regression

Now let us make it even more interesting (see Figure 1-1): say the prices were 160K, 300K, and 440K for houses with one, two, and three bedrooms, respectively. Now, how would you predict the price of a four-bedroom house? One technique you could use is to plot these values on the y axis, with the number of bedrooms on the x axis, and fit a straight line passing through all these points. If no such line exists (i.e., the points are not collinear), then pass a line that passes as closely as possible through these points. This technique of predicting the price using the number of bedrooms, assuming linear relationships between them, is called **linear regression** in machine learning. It is the same technique that you used implicitly in the first two examples.

CHAPTER 1 INTRODUCTION TO MACHINE LEARNING

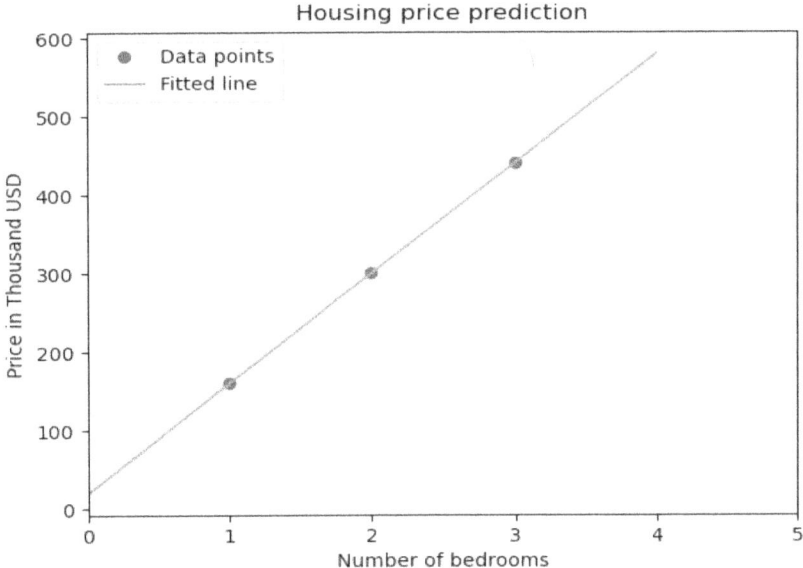

Figure 1-1. *Linear regression*

Mathematically, fitting a line is equivalent to setting variables m (slope) and c (intercept), satisfying the equation y = mx + c. Here, x would be the number of bedrooms in the house, and y would be the price of the house. For the three houses, we have these three equations:

$$160 = m + c$$
$$300 = 2m + c$$
$$440 = 3m + c$$

Solving this gives m = 140 and c = 20. So, the answer to our original question comes out to be 4 * 140 + 20 = 580K, as you can also verify from the line we plotted. In this case, we had three equations in two variables, and we got lucky to find values of m and c that satisfy these three perfectly (because this system with more equations than variables could have very easily been insatiable). However, in general, we cannot rely on luck every time, and therefore, we try to fit a line *as closely* as possible (as opposed to an exact fit, which may not be possible). More precisely, we aim to minimize the **error** as we select these **parameters** m and c. For linear regression, we usually use the mean squared error (MSE), defined as

$$MSE = \frac{1}{n} \sum_{i=1}^{n} (y_i - \hat{y}_i)^2$$

Equation 1

3

where n is the number of points in our set, y_i is the actual output value (i.e., house price), and $\hat{y}_i$ is our predicted price for that point. For other use cases, other formulations of errors can also be used, for example, the mean absolute error (MAE), etc.

The equation y = mx + c can now be used to predict the price of any house given its number of bedrooms (even though the prediction may be highly inaccurate). This model uses only one piece of information as input, i.e., the number of bedrooms, to predict the output. This variable (number of bedrooms) is called a **feature** or an **input variable** of the model.

Every house we considered (i.e., the one, two, and three-bedroom houses) is called a data point, and the collection of these data points we used is called our **input dataset** or **training dataset**. The house for which we made our prediction is called the **test data** – note that this could have comprised more than one example. The variable which we predict, i.e., the price of the house, is called the **target** or **output variable** or **dependent variable** and is usually denoted by y. The rest of the features (also called **independent variables**) are usually denoted by x.

Multivariate Linear Regression

In this example, we used only one feature to predict the house price. In reality, the price of a house may depend on multiple factors. So, we would want to use as many relevant features as possible so that we may get more accurate predictions. We can use, for example, features like the total area in sq. ft., years since construction, garage area, etc. Take a look at this Kaggle competition (https://www.kaggle.com/competitions/house-prices-advanced-regression-techniques), which uses many more features for the same problem.

Now consider a model which uses the following three features, which we denote by a, b, c:

1. **# Bedrooms**: a
2. **Years since construction**: b
3. **Total square footage**: c

The target variable is the house price again, which we denote by y.

Just like the previous example with one feature, we write y as a linear expression of the three input features:

$$y = w_1 * a + w_2 * b + w_3 * c + w_4$$

Here, w_1, w_2, w_3, w_4 are the **model parameters** that we need to select so that the above equation predicts y as accurately as possible. In ML parlance, we say that the ML model needs to **learn** these parameters using what is called the **model training (or learning) algorithm**. One commonly used algorithm for learning unknown parameters in linear regression is **gradient descent**.

Gradient Descent

The primary objective of gradient descent is to learn parameter values that minimize the loss function (i.e., error) over the training dataset. The loss function is an objective function defined in terms of the unknown parameters we aim to learn. Recall that in the case of the previously explained linear regression formulation, we have chosen mean squared error (MSE) as the loss function.

Gradient descent achieves this minimization by calculating the partial derivative of the loss function with respect to each of the learnable parameters, e.g., $\frac{\partial loss}{\partial w_1}$, $\frac{\partial loss}{\partial w_2}$, etc. The derivative $\frac{\partial loss}{\partial w_1}$ gives us the direction in which modifying w_1 causes the greatest increase in loss. Therefore, changing w_1 in the opposite direction, i.e., $-\frac{\partial loss}{\partial w_1}$, would give us the direction of the steepest reduction of loss in terms of w_1. In gradient descent, these gradients are computed at each of the input data points, and the gradients (after multiplication by learning rate) are then subtracted from the parameter values to get new parameter values. This process is repeated for a sufficient number of iterations or until the change in loss between two successive epochs falls below a threshold. The algorithm is then said to have converged to an optimum. We will be using the gradient descent algorithm for model training in most of our projects, and we shall go into more details of this algorithm in the "Model Training" section of this chapter.

Feed-Forward Neural Networks

Linear regression is a simple, yet powerful, ML technique for a variety of problems. However, the dependence of y on the inputs may not always be linear. One way to capture a possible nonlinear dependence is to apply a nonlinear function, also called an **activation function**, after the linear combination of inputs, as $y = g(w_1 * a + w_2 * b + w_3 * c + w_4)$, where g is a nonlinear function (see Figure 1-2). Some of the commonly used activation functions include a rectified linear unit (ReLU), sigmoid, tanh, etc.

CHAPTER 1 INTRODUCTION TO MACHINE LEARNING

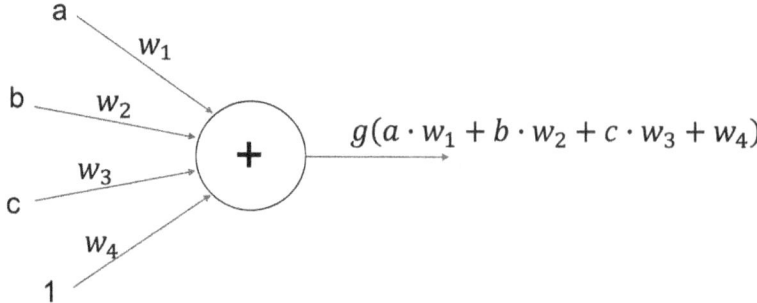

Figure 1-2. A single neuron in a neural network

ReLU or rectified linear unit function (see Figure 1-3) is defined mathematically as ReLU(x) = max(0, x). In other words, whenever x is negative, the function takes on the value of 0. If x is positive, the value of ReLU is simply x.

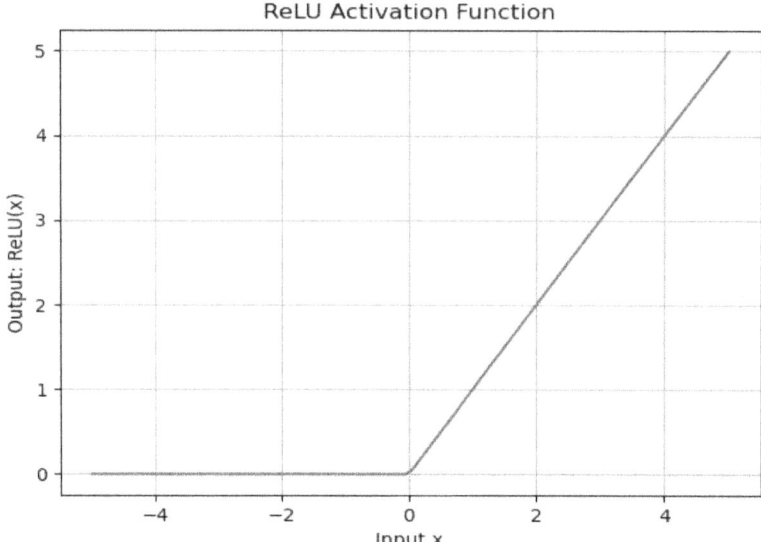

Figure 1-3. ReLU activation function

We shall use ReLU activations in most of our architectures. Some other popular activations include the sigmoid and tanh functions, which are plotted here (Figure 1-4).

CHAPTER 1 INTRODUCTION TO MACHINE LEARNING

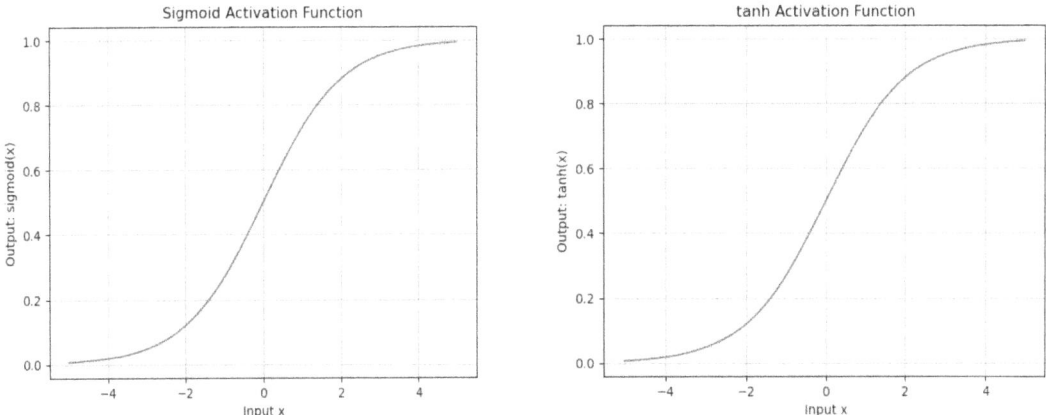

Figure 1-4. *Sigmoid and tanh activation functions*

This unit, which takes a linear combination of inputs followed by a nonlinearity, is also called a **neuron**, and a whole bunch of them connected together in consecutive layers constitute a feed-forward **neural network** (see Figure 1-5).

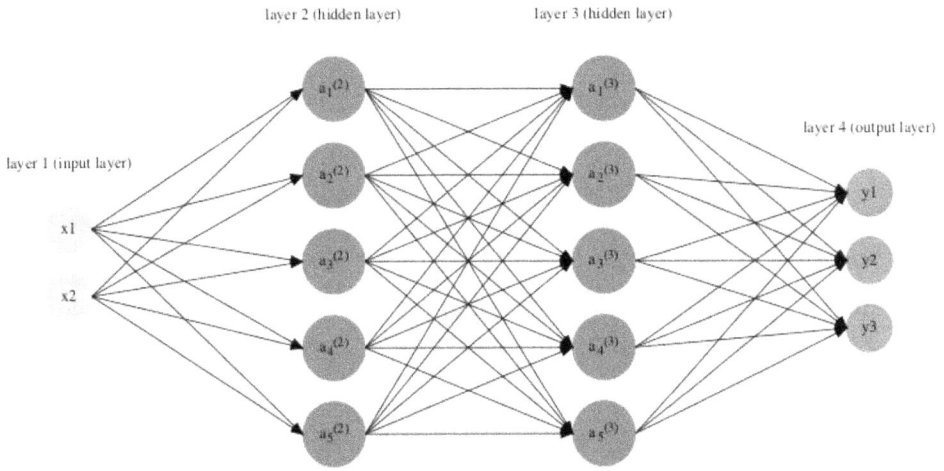

Figure 1-5. *Feed-forward neural network*

From Figure 1-5, notice that the input features constitute the first layer. There are two more layers in this example (there could be many more) called **hidden layers**. Finally, there is the **output layer,** which gives out the prediction or output, which can then be compared to the target (or ground truth) to calculate the loss function. There are connections going from every neuron in a layer to every neuron in the next consecutive layer. Each neuron computes the linear combinations of its inputs, followed by a nonlinear activation function, as explained before.

Machine Learning Primer

We went over some of the ML preliminaries in the previous section informally with the help of the running example of house price prediction. We now go over all these concepts more formally, along with a few additional concepts. Note that we will be working in the domain of supervised learning, where the machine learning model learns from labeled examples. So, whenever we mention ML henceforth, we will mean supervised machine learning.

As we saw, in a typical machine learning setting, we are given a training set of n samples, where each sample contains a <features, label> pair which we typically denote as (x, y); note that here $x = <x_1, x_2, ..., x_m>$ is a vector of features to be used for training. The objective of the training algorithm is to use these training samples to learn the parameters of an ML *model*, which can then be used to predict labels for unseen examples.

There are two types of features that could be used:

1. **Numerical**: These could be floating-point numbers or integers.

2. **Categorical**: Categorical features are those that take values from a fixed set of categories, e.g., gender, taking values from the set {M, F}, or blood group, taking values from the set {A, B, AB, O}, etc.

Similarly, the target variable y could either be a continuous numerical value or it could belong to a finite set of classes. In the former case, we call the problem a **regression task**, whereas in the latter case, we call it a **classification task**. More specifically, if the target y belongs to the set {0, 1}, it is called a binary classification problem.

Train-Test-Validation Split

We typically split all the labelled data we have into three groups: training, validation, and testing. For example, one commonly used ratio is to set aside 10% of the data for testing, 80% of the data for training, and 10% for validation.

The model is typically trained using only the **training data**.

The **test data** is used for the final evaluation of the model, and the metrics thus evaluated are the ones that are reported. The test data is kept hidden from the model during the training process and is only exposed to the final version of the model. This is so that the training algorithm cannot cheat by, for example, memorizing the labels of the test data and regurgitating them from memory. We want our model to be general enough so that it performs well when it sees completely fresh data for the first time. We simulate this "freshness" by hiding the test data from the training algorithm.

While the test data is hidden, how will the training process evaluate its model? For example, we might train several different types of models (e.g., gradient-boosted trees vs. neural networks) and aim to select the better of the two for our use case. For this process, called model selection, we keep aside some data as **validation data**. Validation data is also useful for the process of hyperparameter tuning. This can be achieved by training the model with different settings of hyperparameters (like the max-depth or the number of trees, for the gradient-boosted trees model) and evaluating these trained models on the validation data to select the best setting of hyperparameters. Thus, in a way, the validation dataset acts as a dummy test set to compensate for the unavailability of the test set during training.

Once model selection and hyperparameter tuning are done, it is not uncommon to train the selected model with the selected hyperparameters from scratch using the training as well as the validation data.

We can also use validation data for early stopping. While training the model, we can periodically evaluate it on the validation data. If the validation metrics stop showing significant improvement between consecutive evaluations (or a few consecutive evaluations), you can stop the model training process early. This saves some precious training time and also avoids overfitting.

The train test validation split might require a stratification strategy in the case of imbalanced classification datasets to ensure a similar proportion of labels end up in each split. For example, in the classic task of credit card fraud detection, say only 1% of the transactions are fraudulent. If the train-test-validation split is done without being mindful of the class imbalance, we might end up with only a handful of fraudulent transactions in the training set (say, for example, the training set might contain only 0.1% fraudulent transactions), or vice versa. So we need to instead use a stratified splitting strategy, which would ensure that all the splits would contain the same proportion (1% in this case) of fraudulent transactions.

CHAPTER 1 INTRODUCTION TO MACHINE LEARNING

Figure 1-6 summarizes the high-level ML workflow discussed above.

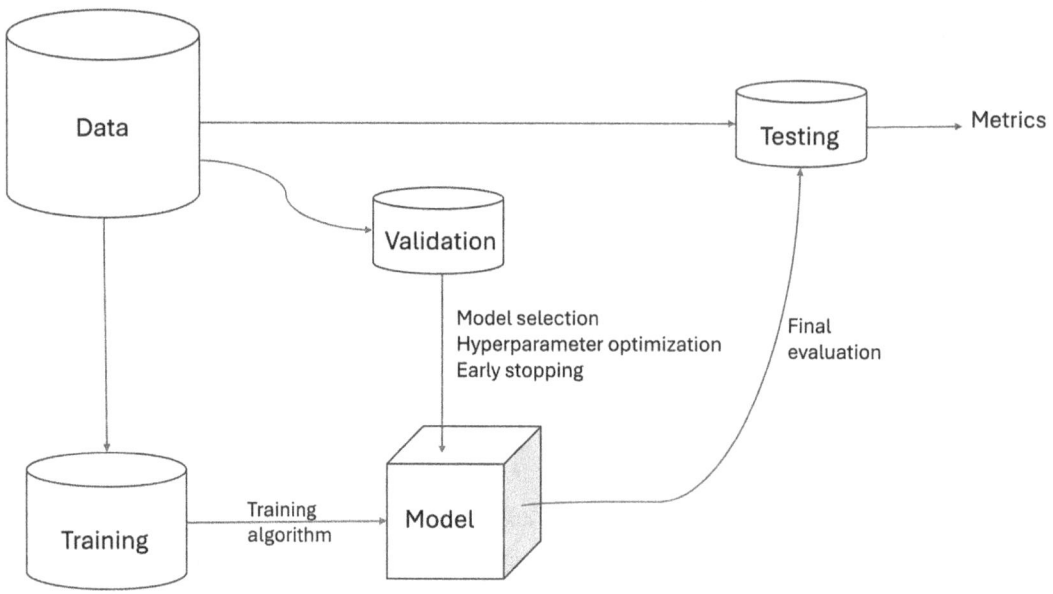

Figure 1-6. *ML workflow*

Model Evaluation

Once the final model is ready, we evaluate it on the test data using some preselected metrics. Depending on the type of the task, we may use different metrics to evaluate it:

1. **Classification Task**: For simplicity, we assume a binary classification task, i.e., the labels belong to the set {0, 1}. Here are some of the basic metrics commonly used for the case of binary classification:

    a. **Accuracy**: This measures the percentage of data points that get labelled correctly by the model.

    b. **Error**: The complement of accuracy, which measures the percentage of data points that get mislabeled by the model.

    c. **Precision and Recall**: For binary classification, there are four scenarios possible depending on whether the target value is actually positive (i.e., label 1) or negative (label 0) and whether the model predicted positive or negative:

Target predicted	Predicted positive	Predicted negative
Actual positive	True positive (TP)	False negative (FN)
Actual negative	False positive (FP)	True negative (TN)

We then define precision and recall as

$$Precision = \frac{TP}{TP + FP}$$

$$Recall = \frac{TP}{TP + FN}$$

Based on these, the F1 score is defined as the harmonic mean of precision and recall:

$$F_1 = 2 \frac{precision \cdot recall}{precision + recall}$$

2. **Regression Task**

   a. **MSE**: The mean squared error, as we saw, is defined as the mean of the squared error of the target vs. the predicted value.

   b. **RMSE**: This takes the square root of the MSE defined above.

   c. **MAE**: Mean absolute error takes the mean of the absolute values of the errors.

Domains of Machine Learning

Based on the techniques used, (supervised) machine learning comprises mainly three sub-domains of problems:

1. **Computer Vision (CV)**: CV deals with image datasets and includes problems such as image classification, object detection, image generation, video analysis, etc.

2. **Natural Language Processing (NLP)**: NLP deals with text-based datasets and includes problems like sentiment analysis, text summarization, machine translation, etc.

3. **Tabular Data**: Traditional ML deals with many problems involving *structured* or *tabular* data, which essentially means data that can be stored in a "table," each row corresponding to a data point, with well-defined columns. Contrast this with a text corpus or an image, which falls under the category of *unstructured* data.

Data Collection

High-quality data is foundational for building robust and accurate ML models. So, the first step in any ML project is collecting a sufficient amount of good-quality data using correct techniques. Big companies have ample data readily available for training huge models. For personal small- and medium-scale projects, you might find that the web is a good data source, especially for tasks related to images or text. For example, even many of the large language models (LLMs) trained by big tech companies use Wikipedia as one of the important sources of training data. Apart from that, there are other sources like Reddit, social media like X (formerly Twitter), etc.

There are also many publicly available datasets for exploration hosted by websites like Kaggle (https://www.kaggle.com/datasets), the UCI ML repository (https://archive.ics.uci.edu/), GitHub-hosted *awesome public datasets (https://github.com/awesomedata/awesome-public-datasets)*, and even some government portals like data.gov (https://data.gov/) by the US government.

In case you collect your own data, labeling is often a challenge that might require some kinds of crowdsourcing techniques.

Data Preprocessing

Raw data might require different types of preprocessing depending on the type of data at hand. Some examples include the following.

Noisy or Missing Data Correction: Some data points might have some features missing; this can be corrected by either simply inserting some default values (e.g., 0 or –1 for numerical features); by imputation, i.e., filling the missing values by the corresponding column mean; or by other suitable statistical techniques. There might also be some obvious noisy values, e.g., an age column containing negative values, or some categorical variable containing values not within the categories, etc. This can also be corrected by mean imputation.

Outliers: Some data points, while being perfectly valid, may contain extreme target values. Using such values in the training data might throw off the model. We can filter out outliers by techniques like removing all data points that have a target value more than some (2 or 3) standard deviations away from the mean target value. Note that this must be done only for the training data and not for the test data, since we still want to evaluate the model on such outliers to get a realistic sense of the model's performance. Also, note that sometimes we might want to keep the outliers since the model might be able to learn from this data to predict correctly for outliers – this decision can be made based on the performance of both methods on the validation data.

Preprocessing in Natural Language Processing: In NLP, a series of preprocessing techniques are usually used, such as stemming, lemmatization, lowercasing, punctuation removal, stop word removal, tokenization, etc. We shall discuss some of these in greater detail in Chapters 4, 5, and 6, when we work on NLP projects.

Preprocessing in Computer Vision: In CV, preprocessing typically involves image resizing, normalization, noise reduction, contrast enhancement, etc. Chapters 3 and 9 will discuss more details on these.

Feature Engineering

Consider another example of a linear regression problem, like in the "Machine Learning with an Example" section, but this time, let us assume that the dependence of y on the input variable x is as shown in Figure 1-7.

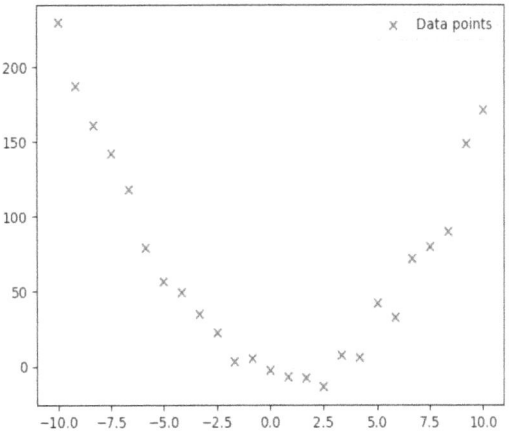

Figure 1-7. Dependence of target (y) on input variable (x)

It is clear that no straight line can model this dependence very well. However, allowing the model to contain terms quadratic in x will enable a good fit, as seen in Figure 1-8.

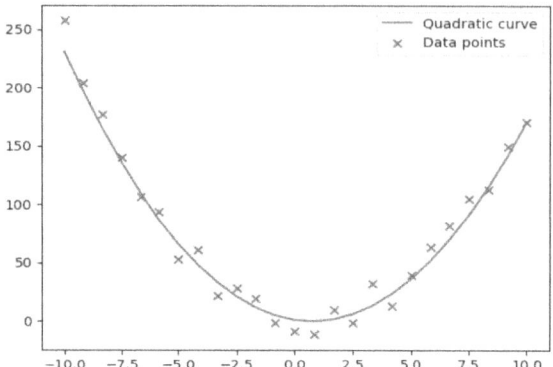

Figure 1-8. *Modeling y as a quadratic in x*

In other words, modeling y as a linear combination of x and x^2 gives us a pretty accurate model. This would involve an equation like this: $y = ax^2 + bx + c$. Note that this is still considered as linear regression because, although it uses features nonlinear in the input x, the model is still linear with respect to the learnable parameters a, b, and c.

More generally, if the input contains features x_1, x_2, …, then instead of just using these as is, we can add higher degree features as well, like $x_1 x_2$, x_1^2, $x_2 x_3 x_4$, etc., if that helps our model. In principle, we can go on adding higher-degree features of this type. However, this is a delicate process because we need to ensure that we don't add so many features that it causes the model to overfit (described in detail in the last section of this chapter).

As you may have realized, feature engineering is an intensive process. It is here that neural networks come to our rescue. On a high level, neural networks extract the features important for our learning task on their own, and this saves us a lot of effort in feature engineering. For example, convolutional neural networks (CNNs) work with images and slowly extract increasingly complex features in each subsequent layer. So, the first few layers might simply detect edges, boundaries, etc. The next few layers might extract more detailed features involving body parts like eyes, hands, or legs. The final layers use these features to detect whether the image contains a cat, a dog, or a human. You will get a much better understanding of how CNNs achieve this in Chapter 3 when we work on an image classification project.

Model Training

Model training involves running the learning algorithm that takes the training data as input and *trains* different parameters of the model. Different training algorithms are used based on the type of model being trained. Neural networks are trained using the popular gradient descent algorithm, which we will now describe in detail since we shall use it frequently throughout this book.

Training Neural Networks Using Gradient Descent and Backpropagation

We shall describe this process of training neural networks in two main steps:

1. How to update parameters using gradients
2. How to calculate the gradients

Let us use the following simple neural network as a running example (see Figure 1-9). It consists of two inputs x_1, x_2 and three neurons labeled as g1, g2, and g3. The weights are denoted as w_1, w_2, ..., w_6. Each neuron applies the RELU activation function on the linear combination of its inputs. The output of neuron g1 (also called activation) is denoted as a_5, and that of g2 is denoted a_6. The output of g3, which is also the output of our neural network, is denoted y. The ground truth label is $\hat{y}$. The final loss is then computed by some loss function.

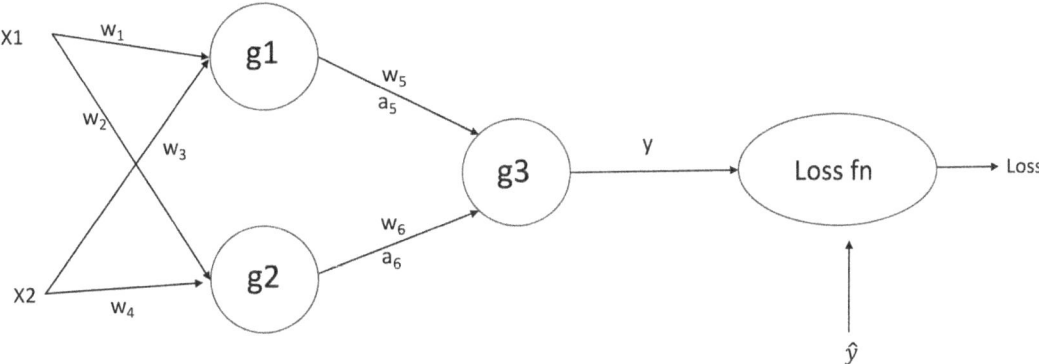

Figure 1-9. Example neural network

Updating Parameters

We will use the MSE loss over n training samples, defined as

$$Loss = \frac{1}{n}\sum_{i=1}^{n}\left(y_i - \widehat{y}_i\right)^2$$

Here, $\widehat{y}_i$ is the target value for the ith sample, whereas y_i is the output predicted by the model for the same.

For the neural network, the only learnable parameters are the weights $w_1, w_2, ...w_6$, and these are the parameters that our training algorithm (i.e., gradient descent) will aim to learn. The algorithm will first initialize all these weights randomly.

This algorithm will work in two stages:

1. **Forward Pass**: During this phase, the algorithm will pass all the inputs from the training data through the model and compute the MSE loss as defined in the equation above.

2. **Backward Pass**: This is the stage where all the parameters will be updated according to the gradients as

$$w_i = w_i - \alpha \frac{\partial Loss}{\partial w_i}$$

Here, α is a hyperparameter called the **learning rate**. Let us say that it is set to 0.1 for now. During the backward pass, the gradients *flow* backward from the output toward the input using what is called the chain rule for derivatives. We shall now describe this process of gradient computation during the backward pass.

The algorithm undergoes several iterations, alternating between the forward pass and the backward pass: in each iteration, it estimates the loss using the current parameter values during the forward pass, and based on this loss, it updates the parameter values (using gradients) in the backward pass.

Some other variants of gradient descent are also commonly used, for example, mini-batch gradient descent passes a batch of data in each iteration instead of passing the entire training dataset. For all these algorithms, a passage of the entire training data through the model (i.e., forward and backward pass) is called an epoch.

Gradient Computation

A gradient is computed in neural networks through the process of backpropagation, which is based mainly on the chain rule in calculus, which goes as

$$\frac{\partial z}{\partial x} = \frac{\partial z}{\partial y} * \frac{\partial y}{\partial x}$$

This essentially says that the partial derivative of z with respect to a variable x is the product of the partial derivative of z with respect to an intermediate variable y, multiplied by the partial derivative of this variable y with respect to x.

We now take the partial derivative of the loss with respect to y_i, the target for sample i:

$$\frac{\partial Loss}{\partial y_i} = \frac{2}{n}\left(y_i - \widehat{y}_i\right)$$

Note that all the terms in the summation vanish except the one corresponding to sample i, since only that term depends on y_i (we assume all the samples to be independent).

Also, we have $y_i = ReLU(a_{i5}w_5 + a_{i6}w_6)$. Here, we are denoting the activation from neuron g1 (in Figure 1-8) for the ith sample by a_{i5} and that from neuron g2 by a_{i6}. Let us assume that ReLU is working in the activated, linear regime (i.e., its input is positive). In that case, $y_i = a_{i5}w_5 + a_{i6}w_6$.

Therefore, $\frac{\partial y_i}{\partial w_6} = a_{i6}$. Combining these, we can now calculate the partial derivative of the loss with respect to w_6 using the chain rule as

$$\frac{\partial Loss}{\partial w_6} = \frac{\partial Loss}{\partial y_i} * \frac{\partial y_i}{\partial w_6} = \frac{2}{n}\left(y_i - \widehat{y}_i\right) \bullet a_{i6}$$

where we used expressions for $\frac{\partial Loss}{\partial y_i}$ and $\frac{\partial y_i}{\partial w_6}$ from our calculations before. Analogously, the partial derivative with respect to w_5 can be seen to be $\frac{2}{n}\left(y_i - \widehat{y}_i\right) \bullet a_{i5}$.

Similarly, we can calculate the partial derivative of loss with respect to a_{i5} as

$$\frac{\partial Loss}{\partial a_{i5}} = \frac{\partial Loss}{\partial y_i} * \frac{\partial y_i}{\partial a_{i5}} = \frac{2}{n}\left(y_i - \widehat{y}_i\right) \bullet w_5$$

CHAPTER 1 INTRODUCTION TO MACHINE LEARNING

Similarly, the partial derivative with respect to a_{i6} can be seen to be $\frac{2}{n}(y_i - \widehat{y_i}) \bullet w_6$.

It is clear that we need $\frac{\partial Loss}{\partial w_5}, \frac{\partial Loss}{\partial w_6}$ for updating the parameters w_5, w_6. But you might wonder: Why do we need to compute $\frac{\partial Loss}{\partial a_{i5}}, \frac{\partial Loss}{\partial a_{i6}}$? This will be clear next:

To calculate the gradients for parameters from the first layer, we can again use the chain rule:

$$\frac{\partial Loss}{\partial w_1} = \frac{\partial Loss}{\partial a_{i5}} * \frac{\partial a_{i5}}{\partial w_1}$$

Now, we have already computed $\frac{\partial Loss}{\partial a_{i5}}$.

Also, $a_{i5} = RELU(w_1 x_{i1} + w_3 x_{i2}) = w_1 x_{i1} + w_3 x_{i2}$, assuming we are in the linear regime for RELU. So, we have: $\frac{\partial a_{i5}}{\partial w_1} = x_{i1}$.

Combining these two,

$$\frac{\partial Loss}{\partial w_1} = \frac{2}{n}(y_i - \widehat{y_i}) \bullet w_5 \bullet x_{i1}$$

Notice that we could calculate the gradient with respect to w_1 using gradient of loss with respect to a_{i5} (backpropagated from the previous step) and the local gradient of a_{i5} with respect to w_1. Therefore, for the second layer, we needed the gradients for the weights as well as those for the inputs to that layer, to be able to apply the chain rule. So, in general, we want to calculate the gradients for both the weights (i.e., parameters) and the inputs (i.e., activations) for all the layers (except for the first layer, since there is no layer preceding the first layer to which the gradients would need to flow).

Also, to calculate the gradients of the loss with respect to parameters in any layer, we just need to multiply the gradients of loss with respect to the outputs of this layer (i.e., inputs of the next layer) by the local gradients of the outputs with respect to the parameters. We can analogously also calculate the gradients of the loss with respect to the activations (i.e., outputs) of any layer.

In this way, the gradients can be thought of as flowing backward. Therefore, gradients are computed in the backward direction from the final output layer to the input layer, and this phase is called the backpropagation of gradients.

CHAPTER 1 INTRODUCTION TO MACHINE LEARNING

Overfitting and Underfitting

Figure 1-10 plots the feature values of some sample points on the x axis and the corresponding target labels on the y axis. The solid curve fits the data points very closely – it passes through almost every one of the 25 data points.

On the other hand, the dashed curve, while erring a little on some of the data points, still seems to fit the data reasonably well while also being smoother.

Which of these two curves seems to be more "plausible," or in other words, which one do you think is more likely to be representative of the data? Or, more practically, given a new input x, which of these would you bet on to predict y better?

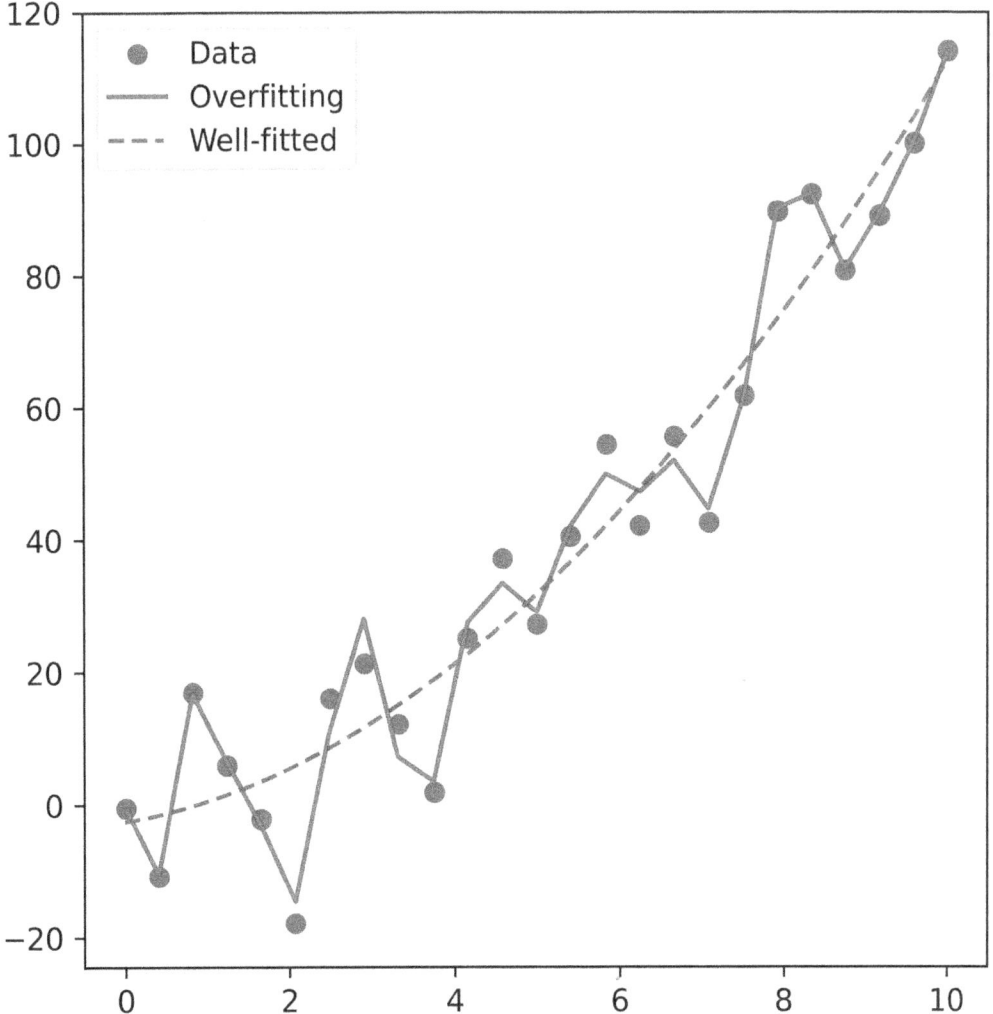

Figure 1-10. *Overfitting example*

I think you would agree that the dashed curve is a better bet at predicting the data. In general, any data that we observe is bound to be somewhat noisy. If we try to fit our ML model to the training data perfectly, the model will also end up learning this "noise" instead of learning the real "signal" or the underlying pattern. Therefore, it won't be able to generalize well when it sees new data. This phenomenon of fitting the training data too closely, while losing generalizability, is called **overfitting**.

In general, if a model becomes overly "complex," it is more likely to overfit the training data. To avoid this, we use techniques to penalize model complexity. This method is called **regularization**. For example, in linear regression, a common practice is to add *regularization* terms to the loss function, which could either be absolute values of the coefficients (as in L1 regularization) or the squares of the coefficients (as in L2 regularization). Both these methods assist in reducing the magnitude of the coefficients in linear regression.

On the flip side, a model that is too simple will not capture much of the noise, but it will not be able to capture the pattern in the data either. In the example above, an extremely simple model might only predict a constant value irrespective of x. Such a model would be practically useless since it does not make use of the training data at all. This phenomenon, where the patterns in training data are not sufficiently captured by the model, is called **underfitting**.

Summary

- Artificial intelligence focuses on developing humanlike cognitive abilities within computers. It often involves using machine learning, which is a set of data-driven techniques to model different phenomena.

- Linear regression is an ML technique that involves trying to model the dependent variable y as a linear combination of input features x_i. The coefficients of this linear dependence are the parameters of the model, which we can train using the gradient descent algorithm.

- Feed-forward neural networks involve neurons connected to each other in a sequence of layers. Each neuron computes a linear combination of its inputs, followed by applying a nonlinear activation function such as ReLU.

- Data samples in supervised learning comprise features, which could be numerical or categorical, and a target label, which could be a continuous numerical value (regression task) or could belong to a fixed set of classes (classification task).

- It is common to partition data in an ML project into three parts: training data, used for model training; validation data, used for hyperparameter tuning, model selection, or early stopping; and testing data, used for the final model evaluation.

- Different metrics can be used for model evaluation depending on the task (classification vs. regression).

- Data collection is a crucial component of any ML project, and so is data preprocessing, which involves steps like handling missing values, noisy data, or outliers, and more domain-specific techniques for areas like computer vision or natural language processing.

- Feature engineering is an important and challenging task, which might involve adding higher-degree terms of input variables as features. Feature extraction happens automatically in neural networks, which is one of their big advantages.

- Gradient descent is a popular algorithm used in model training. It involves repeatedly passing the training data through the model and updating the model parameters in each iteration via backpropagating the gradients.

- A model that is too complex runs the risk of overfitting on the training data, making it less generalizable on fresh data. Regularization methods help to avoid this problem of overfitting. On the other hand, a model that is too simple risks causing underfitting.

CHAPTER 2

Tensors in PyTorch

This chapter introduces the reader to tensors, the foundational data structure of PyTorch. We start the chapter by introducing tensors in PyTorch, building from the simplest 1D tensors to 2D and then 3D tensors. We gradually build intuition by fixing one dimension at a time and then introduce basic operations like indexing and slicing. We give a glimpse of how tensors are stored in memory. Then we look at different ways of initializing tensors and describe the attributes of tensors. We give a bird's-eye view of tensor operations belonging to different categories to make the reader aware of all that PyTorch has to offer in terms of tensor operations. We then describe how PyTorch's autograd mechanism computes gradients by constructing computational graphs and also go over the high-level recipe that we typically use for model training by leveraging PyTorch's autograd mechanism. We end the chapter with a series of exercises designed to improve your understanding of PyTorch tensors.

Introduction to Tensors

Tensors are the primary data structure used in PyTorch. Tensors are essentially multidimensional arrays with a uniform data type. If you are familiar with NumPy, you will find that PyTorch tensors are very similar to NumPy ndarrays. However, we will assume no prior knowledge of NumPy and introduce PyTorch tensors from scratch. Our approach will be to introduce tensors gradually using examples. We will first define one-dimensional tensors and then 2D tensors and then build up different concepts as we go.

CHAPTER 2   TENSORS IN PYTORCH

In PyTorch, a **one-dimensional tensor** can be constructed from a list as follows.

```
[11]: one_dim_tensor = torch.tensor([1,2,3])
      one_dim_tensor
```

```
[11]: tensor([1, 2, 3])
```

```
[12]: one_dim_tensor.size()
```

```
[12]: torch.Size([3])
```

```
[13]: one_dim_tensor.shape
```

```
[13]: torch.Size([3])
```

You can view its size by using the size() method or the shape attribute. Notice that a 1D tensor can be thought of as a list, with the additional functionalities that come with a tensor (as we will see later).

A **2D tensor** is essentially a matrix, as illustrated here:

```
[17]: two_dim_tensor = torch.tensor([[1,2,3],[4,5,6]])
      two_dim_tensor
```

```
[17]: tensor([[1, 2, 3],
              [4, 5, 6]])
```

```
[18]: two_dim_tensor.size()
```

```
[18]: torch.Size([2, 3])
```

Let us now put this 2D tensor under the scanner and see what happens if we fix one dimension at a time. We can fix dimension 0 to the value "0" by saying `two_dim_tensor[0,:]`. This corresponds to selecting the 0th row in the matrix or, in other words, the 0th element of dimension 0. The colon ":" symbol after the comma signifies that we select all the elements of dimension 1.

```
two_dim_tensor[0, :] # Zeroeth row
```
```
tensor([1, 2, 3])
```

Similarly, let's fix dimension 1 of the tensor to value "2", which will give us the second column of the tensor.

```
two_dim_tensor[:, 2] # Second column
```
```
tensor([3, 6])
```

CHAPTER 2   TENSORS IN PYTORCH

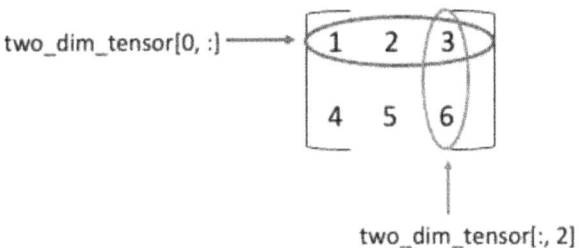

We now look at the first two rows of the tensor. Notice that it has only two rows and trying to select the third row throws an **IndexError**.

```
two_dim_tensor[0, :] # Zeroeth row

tensor([1, 2, 3])

two_dim_tensor[1, :] # First row

tensor([4, 5, 6])

two_dim_tensor[2, :] # Second row
```

```
IndexError                     Traceback (most recent call last)
Cell In[41], line 1
----> 1 two_dim_tensor[2, :]

IndexError: index 2 is out of bounds for dimension 0 with size 2
```

It is worth noting the names of these errors, and we will try to go over a few of them during this book, so that when you see "IndexError" while debugging your code, you will realize immediately that you must be trying to index a dimension out of its bounds. Here, the error gives further details that you tried to index with value "2", which was out of bounds since dimension 0 had a size of only "2".

Similarly, here are all three columns of our matrix, selected one at a time:

```
two_dim_tensor[:, 0] # Zeroeth column

tensor([1, 4])

two_dim_tensor[:, 1] # First column

tensor([2, 5])

two_dim_tensor[:, 2] # Second column

tensor([3, 6])
```

25

CHAPTER 2   TENSORS IN PYTORCH

Now, let us go further and look at a **3D tensor**:

```
[25]: three_dim_tensor = torch.tensor([[[1,2],[3,4]], [[5,6],[7,8]], [[9,10],[11,12]]])
      three_dim_tensor
[25]: tensor([[[ 1,  2],
               [ 3,  4]],

              [[ 5,  6],
               [ 7,  8]],

              [[ 9, 10],
               [11, 12]]])
[26]: three_dim_tensor.size()
[26]: torch.Size([3, 2, 2])
```

Once again, fixing dimensions will help build our intuition of 3D tensors. We can fix dimension 0 to the value "0" by the line `three_dim_tensor[0,:,:]`. This corresponds to selecting the 0th row (or 0th element, if you will) in dimension 0. This will give a tensor of shape [2,2], since the 0th dimension is already fixed to 0 and the only degrees of freedom you have are in selecting the first and second dimensions. This slice of the tensor will contain all the elements belonging to the 0th row of dimension 0.

```
three_dim_tensor[0,:,:] # Fixing dimension 0 to value "0"

tensor([[1, 2],
        [3, 4]])

three_dim_tensor[1,:,:] # Fixing dimension 0 to value "1"

tensor([[5, 6],
        [7, 8]])

three_dim_tensor[2,:,:] # Fixing dimension 0 to value "2"

tensor([[ 9, 10],
        [11, 12]])
```

Let us see what happens when we fix dimension 1 now.

```
three_dim_tensor[:,0,:] # Fixing dimension 1 to value "0"

tensor([[ 1,  2],
        [ 5,  6],
        [ 9, 10]])

three_dim_tensor[:,1,:] # Fixing dimension 1 to value "1"

tensor([[ 3,  4],
        [ 7,  8],
        [11, 12]])
```

So essentially, fixing each dimension of a 3D tensor yields a matrix in the remaining dimensions.

You can also talk about the coordinates (or the "address") of each element: it is the tuple containing the position of that element in each dimension. For example, the very first element 1 belongs to the 0th element of dimension 0, the 0th element of dimension 1, and the 0th element of dimension 2. So, it has coordinates [0,0,0].

Q.: What is the element with coordinates [1,1,1]?

Answer: 8

Note that using indexing in this way will reduce the dimension of the tensor. As an example, `three_dim_tensor[:,0,1].size()` gives `torch.Size([3])`. This is because we have indexed dimensions 1 and 2, thereby removing those from the output, to give a one-dimensional tensor.

Note that a 0-dimensional tensor is simply a scalar value in a tensor and can be defined as below:

```
zero_dim_tensor = torch.tensor(5)
zero_dim_tensor.size()
>>> torch.Size([])
```

Slicing

Slicing in PyTorch tensors works similarly to slicing in Python lists.

Let us work with a 1D tensor first:

```
A = torch.arange(10)
print(A)
>>> tensor([0, 1, 2, 3, 4, 5, 6, 7, 8, 9])
```

Now, we can slice the first three elements:

```
A[:3] # First 3 elements
>>> tensor([0, 1, 2])
```

The last three elements:

```
A[7:] # Last three elements
>>> tensor([7, 8, 9])
```

The last three elements again, in a clearer format:

```
A[-3:] # Also, last three elements
>>> tensor([7, 8, 9])
```

This is because the index -1 corresponds to the last element, -2 to the second last element, and so on. This latter way is more readable since it does not require you to calculate the index from which you need to start, to get the last three elements: for example, imagine if A was a 1024 length tensor, then A[961:] does not immediately convey that you selected the last 64 elements, as opposed to the simpler A[-64:].

A slice with the middle four elements:

```
A[3:7] # Middle 4 elements
>>> tensor([3, 4, 5, 6])
```

Slicing with steps of two, selecting every other element:

```
A[0:-1:2] # Alternate elements
>>> tensor([0, 2, 4, 6, 8])
```

Here, the first 0 before the : specifies that we are starting from the 0th element (this was redundant, added for clarity), the -1 after the : specifies that we are going up to the very last element (again redundant), and the final 2 after the : specifies that we are slicing elements in steps of two. Note that A[::2] would have achieved the same.

Let us now revisit the 3D tensor we defined before for slicing:

```
B = torch.tensor([[[1,2],[3,4]], [[5,6],[7,8]], [[9,10],[11,12]]])
B
tensor([[[ 1,  2],
         [ 3,  4]],

        [[ 5,  6],
         [ 7,  8]],

        [[ 9, 10],
         [11, 12]]])
```

Slicing the first two rows of dimension 0 gives the following:

```
B[:2,:,:]
tensor([[[1, 2],
         [3, 4]],

        [[5, 6],
         [7, 8]]])
```

Similarly, slicing along dimension 1 gives

```
B[:,:1,:]
tensor([[[ 1,  2]],

        [[ 5,  6]],

        [[ 9, 10]]])
```

Essentially, this has selected the first row in each of the three 2 × 2 matrices of tensor B.

Slicing along dimension 2 selects the second "column" from each of those 2×2 matrices:

```
B[:,:,1:]
tensor([[[ 2],
         [ 4]],

        [[ 6],
         [ 8]],

        [[10],
         [12]]])
```

Putting all these three together gives an intersection of each of those individual slicings:

```
B[:2,:1,1:]
tensor([[[2]],
        [[6]]])
```

You may have noticed that slicing, unlike indexing, does not remove the corresponding dimensions from the tensor. This is further illustrated by the following snippet:

```
print(f"B has shape: {B.size()}")
print(f"Sliced tensor: {B[:1,:1,:]} has shape: {B[:1,:1,:].shape}")
print(f"Indexed tensor: {B[0,0,:]} has shape: {B[0,0,:].shape}")
>>> B has shape: torch.Size([3, 2, 2])
>>> Sliced tensor: tensor([[[1, 2]]]) has shape: torch.Size([1, 1, 2])
>>> Indexed tensor: tensor([1, 2]) has shape: torch.Size([2])
```

Tensors As Logical Views of Physical Memory

A PyTorch tensor – even if it has a dimension greater than 1 – is stored in a contiguous block in physical memory. For example, the 3D tensor B defined above is stored as a contiguous array of integers from 1 to 12 in physical memory. The tensor B itself simply offers a logical view of this block of memory. The shape of B is used to index this contiguous memory in the appropriate way.

Since B is simply a logical view of memory, reshaping B just gets a new view of this same physical memory:

```
C = B.reshape(2,6)
print(C)
tensor([[ 1,  2,  3,  4,  5,  6],
        [ 7,  8,  9, 10, 11, 12]])
```

Now, if C is modified, it causes a change in the underlying memory and therefore causes a corresponding change in B as well.

```
C[0,:] = 0 # Zeroes out the first row of C
print(B)
tensor([[[ 0,  0],
         [ 0,  0]],

        [[ 0,  0],
         [ 7,  8]],

        [[ 9, 10],
         [11, 12]]])
```

Because of these two properties, i.e., storage in a contiguous memory block and having a uniform data type, it is computationally very efficient to operate on PyTorch tensors. This is in contrast with Python lists, which have neither of these two properties and would therefore be significantly slower for our purposes (although of course lists are a very versatile data structure in Python with diverse applications).

Initializing Tensors

So far, we have seen how to initialize a PyTorch tensor using a list. Let us now look at some other ways of initializing tensors that may be more suitable depending on the use case.

From Another PyTorch Tensor

```
tensor_1 = torch.tensor([[1,2],[3,4]])
tensor_2 = torch.tensor(tensor_1)
tensor_2

tensor([[1, 2],
        [3, 4]])
```

With Constant Values

PyTorch provides special functions to create a tensor filled with all 1s or all 0s. These can be created by specifying the shape as

```
shape = (3,2,)
all_ones_tensor = torch.ones(shape)
all_zeros_tensor = torch.zeros(shape)

all_ones_tensor

tensor([[1., 1.],
        [1., 1.],
        [1., 1.]])

all_zeros_tensor

tensor([[0., 0.],
        [0., 0.],
        [0., 0.]])
```

CHAPTER 2 TENSORS IN PYTORCH

It can also be created to mimic the shape and data type of an existing tensor, like

```
tensor_1 = torch.tensor([[1,2],[3,4]])
all_ones_tensor = torch.ones_like(tensor_1)
all_zeros_tensor = torch.zeros_like(tensor_1)

all_ones_tensor

tensor([[1, 1],
        [1, 1]])

all_zeros_tensor

tensor([[0, 0],
        [0, 0]])
```

From a NumPy ndarray

A torch tensor can be created directly from a NumPy array:

```
import numpy as np
np_array = np.array([[1,2],[3,4]])
torch_tensor = torch.from_numpy(np_array)
torch_tensor

tensor([[1, 2],
        [3, 4]])
```

However, the torch tensor will share the same storage as the NumPy array, so that any change in the NumPy array will also reflect in the torch tensor created from it:

```
np_array[0,0] = 100
torch_tensor

tensor([[100,   2],
        [  3,   4]])
```

Creating Random Tensors

The rand() function creates a tensor with the specified size containing values sampled uniformly from [0,1).

```
random_tensor = torch.rand(2,3) # Tensor with values sampled uniformly from [0,1)
random_tensor

tensor([[0.8266, 0.5354, 0.2155],
        [0.8084, 0.4670, 0.1325]])
```

Similarly, `rand_like()` creates a tensor with the same shape as the input tensor and also the same data type (which is why tensor_1 below had to have dtype float32 and not ints):

```
tensor_1 = torch.tensor([[1.0,2.0],[3.0,4.0]])
random_tensor_2 = torch.rand_like(tensor_1)
random_tensor_2

tensor([[0.0100, 0.0973],
        [0.1087, 0.5465]])
```

PyTorch also offers functionality of sampling from other common distributions, with functions like `bernoulli()`, `multinomial()`, `normal()`, `poisson()`, `randn()`, and `randn_like()`, where most of the names are descriptive enough to give an idea of their functionalities.

The function `randn()` samples from the standard normal distribution (with mean 0, variance 1). As you might have guessed by now, functions with a "_like" suffix create tensors with shape and dtype matching the input tensor. So `randn_like()` samples from the standard normal distribution, so that the output matches the input in shape and dtype.

We also have `randint()` and `randint_like()`, which sample random integers uniformly between a specified low and high, and `randperm()`, which outputs a random permutation of integers between 0 and n-1.

PyTorch also offers functions related to the seeds for random number generation, like seed(), manual_seed(), initial_seed(), and also the states of the random number generator, like get_rng_state() and set_rng_state(). We will explain these functions as we use them in later chapters.

You can find more details on random sampling in the official documentation here (https://docs.pytorch.org/docs/stable/torch.html#random-sampling).

There are many more ways of creating tensors, as you can see in this official documentation (https://docs.pytorch.org/docs/stable/torch.html#tensor-creation-ops). We give an overview of some interesting categories of creation ops here:

1. **Sparse Tensors**: sparse_coo_tensor, sparse_csc_tensor, etc., which create sparse tensors in different formats

2. **Creating from Different Sources**: Like asarray, as_tensor, from_file, frombuffer, etc.

3. **Tensors Containing Systematic Data**: Like range, linspace, logspace, eye, etc.

4. **Quantized Tensors**: Like quantize_per_tensor, quantize_per_channel, dequantize

Tensor Attributes

Each PyTorch tensor has three attributes:

1. **torch.dtype**: The data type of the tensor. Recall that all members of the tensor must belong to the same data type. The primary data types in PyTorch are those representing floating-point numbers (float32, float64, float16, bfloat16), complex numbers (complex64, complex128), integers (uint8, int8, int16, int32, int64), and Booleans (bool). The number next to the type represents the number of bits, e.g., int16 corresponds to 16-bit signed integers, and so on.

2. **torch.device**: The device on which the tensor is stored. This is usually "cpu", unless you specify a GPU device.

3. **torch.layout**: The memory layout of the tensor. This is torch. strided for the more common case of dense tensors (which is what we will use for the rest of this book unless otherwise stated). For storing sparse tensors, there is currently beta support for torch. sparse_coo.

```
tensor_1 = torch.tensor([[1.0, 2.0],[3.0, 4.0]])
print(f'tensor_1.dtype: {tensor_1.dtype}')
print(f'tensor_1.device: {tensor_1.device}')
print(f'tensor_1.layout: {tensor_1.layout}')
```

```
tensor_1.dtype: torch.float32
tensor_1.device: cpu
tensor_1.layout: torch.strided
```

Tensor Operations: A Bird's-Eye View

The official documentation page classifies tensor operations into the following categories, which we further sub-categorize to give the reader a sense of all the operations available in PyTorch.

1. **Pointwise Ops**

    a. **Basic Math Ops**: Includes abs, add, addcdiv, addcmul, ceil, clamp, etc.

    b. **Trigonometric Ops**: acos, acosh, asin, atan, sin, sinh, etc.

    c. **Logarithmic**: log, log10, log2, exp, etc.

    d. **Logical**: logical_or, logical_and, logical_not, etc.

    e. **Others**: sigmoid, softmax, gradient, etc.

2. **Reduction Ops**: These are operations that typically reduce the entire tensor to a scalar value or, more precisely, a 0-dimensional tensor

    a. **Min and Max**: argmax, argmin, max, min, etc.

    b. **Over Boolean Tensors**: any, all

    c. **Statistics**: mean, median, mode, quantile, std, count_nonzero, etc.

3. **Comparison Ops**: Operations compare two tensors:

    a. **Element-Wise Comparisons**: eq, ge, gt, etc.

    b. **Element-Wise Min/Max**: minimum, maximum

    c. **Membership to a Set/Type**: isin, isnan, isreal, isinf, etc.

    d. **Sorting**: sort, topk, msort

4. **Spectral Ops**: stft, istft, hamming_window, etc.

5. **Other Ops**: Apart from these, PyTorch offers several other interesting ops like bincount, broadcast_to, clone, combinations, and many more.

Make sure to glance through the complete list so that you are aware of all that PyTorch has to offer. We will cover some of the most frequently used ops in the set of exercises at the end of this chapter.

Computational Graphs and Gradients

As we saw in Chapter 1, a neural network has some learnable parameters, and the goal of model training is to learn these parameters in such a way as to minimize the loss function. We use the gradient descent or other optimization algorithms for our training process, with a forward pass for computing the output and the loss function and the backward pass for gradient computation (see Chapter 1 for details).

PyTorch's autograd mechanism makes this process of backpropagation very easy for us to implement while also being quite efficient computationally. In PyTorch, when a tensor with requires_grad set to True undergoes any operation, PyTorch adds a gradient function (grad_fn) to the new tensor created as a result of this operation.

Let us see a small example:

```
a = torch.tensor([1.0,2.0], requires_grad=True)
b = torch.tensor([4.0,3.0], requires_grad=True)
```

We first defined two tensors, a and b, setting requires_grad to True so as to track their gradients.

```
c = a.pow(3) + b.square()
print(c)
tensor([17., 17.], grad_fn=<AddBackward0>)
```

We now define the tensor $c = a^3 + b^2$. You may have noticed that, unlike all the other tensors we have seen so far, this one shows not only the data it holds but also the associated grad_fn, which it maintains to compute the gradient during the backward pass.

```
e = c.sum()
print(e)
tensor(34., grad_fn=<SumBackward0>)
```

In PyTorch, the backward() function can only be called on a scalar (i.e., 0-dimensional tensor). We define e to be the sum of the elements of c, which will allow us to call backward() on e:

```
e.backward()
print(f"a.grad: {a.grad}")
print(f"b.grad: {b.grad}")
```

```
a.grad: tensor([ 3., 12.])
b.grad: tensor([8., 6.])
```

As you can verify analytically, if you represent a and b as vectors $a = (a_1, a_2)$, $b = (b_1, b_2)$, the gradient of $e = a_1^3 + b_1^2 + a_2^3 + b_2^2$ with respect to a is $(3a_1^2, 3a_2^2)$, which is (3, 12). Similarly, calculate by hand and verify that b.grad matches the value of the gradient of e with respect to b.

Let us now also see what c.grad and e.grad are:

```
print(f"c.grad: {c.grad}")
print(f"e.grad: {e.grad}")
c.grad: None
e.grad: None
```

This is because during a call to backward(), PyTorch only populates .grad attributes for *leaf tensors,* which also have requires_grad set to True.

The following types of tensors are defined to be leaf tensors:

1. All tensors that have requires_grad = False

2. All tensors that have requires_grad = True, and have been created by the user directly

You can easily check if a tensor is a leaf by using the is_leaf attribute of the tensor. Let us see which of our tensors in the previous examples are leaves:

```
print(f"a.is_leaf : {a.is_leaf}")
print(f"b.is_leaf : {b.is_leaf}")
print(f"c.is_leaf : {c.is_leaf}")
print(f"e.is_leaf : {e.is_leaf}")
a.is_leaf : True
b.is_leaf : True
c.is_leaf : False
e.is_leaf : False
```

As we would expect, a and b are leaves since they have requires_grad set to True and are user-defined. Tensors c and e have requires_grad set to True but are not user-defined; therefore, they are not leaf nodes.

To summarize, a variable has a .grad attribute populated only if

1. It is a leaf, i.e., is_leaf is True for it.

2. Requires_grad is set to True for it.

You can use the retain_grad() function if you wish to have .grad populated for any non-leaf tensor.

As we saw in the example, whenever an operation is performed on a tensor with requires_grad = True, the resultant tensor is attached with a grad_fn. This resultant tensor also has its requires_grad set to True since it was the result of an operation that had at least one input with requires_grad = True. This set of operations on tensors can be seen to define a directed acyclic graph (DAG), also called a computational graph.

A computational graph has operations as nodes. Its edges can be thought of as corresponding to the flow of data (i.e., tensors). Inputs to a computational graph are leaf tensors, whereas outputs are called roots (on which backward is usually called).

When the backward() function is called from any root node with a scalar value, it computes the gradient of each of the functions in the reverse order of the computational graph, multiplying them along the way as per the chain rule.

Gradient Descent in Neural Networks

Let us now see the high-level recipe we typically use to train our neural networks using this autograd mechanism and gradient descent:

1. **Model Definition**: The first step is to define a model as a subclass of the nn.Module class. For now, we shall load a pretrained neural network model from PyTorch's torchvision module for examination; the neural networks that we handcraft later will be similar to this:

    ```
    from torchvision import models
    model = models.resnet18(pretrained=True)
    ```

    The parameters (i.e., weights and biases) of a neural network model can be accessed via the parameters() method. We see here that they are all leaves of the computational graph, with requires_grad set to True:

```
for param in model.parameters():
    assert param.is_leaf
    assert param.requires_grad
    # print(param)
```

Therefore, as we saw before, .grad attributes will be populated for all these parameters during a call to backward().

Go ahead and print out the parameters by a `print(param)` statement in the loop above to get a better sense of them: you will see a long list of tensors comprising the model parameters.

2. **Forward Pass**: We pass our training data (usually in batches) through the model to get the model predictions with something like `predictions = model(train_data)` (don't run this line with any data as yet; this is meant for representation purposes, and we will see examples of this in the next chapter using actual data). The loss function, designed to capture the error between the predicted and the actual label values, is also further computed for these sets of predictions.

3. **Backward Pass**: We call loss.backward(), which then populates the gradients for all our model parameters.

4. The model parameters are updated using the gradients populated in .grad attributes as per the gradient descent algorithm we described in Chapter 1 (see Figure 2-1). This process is much easier than it sounds since there are out-of-the-box optimizers in the torch.optim library that take these parameters as inputs, along with hyperparameters like learning rate, and update the parameters in each iteration.

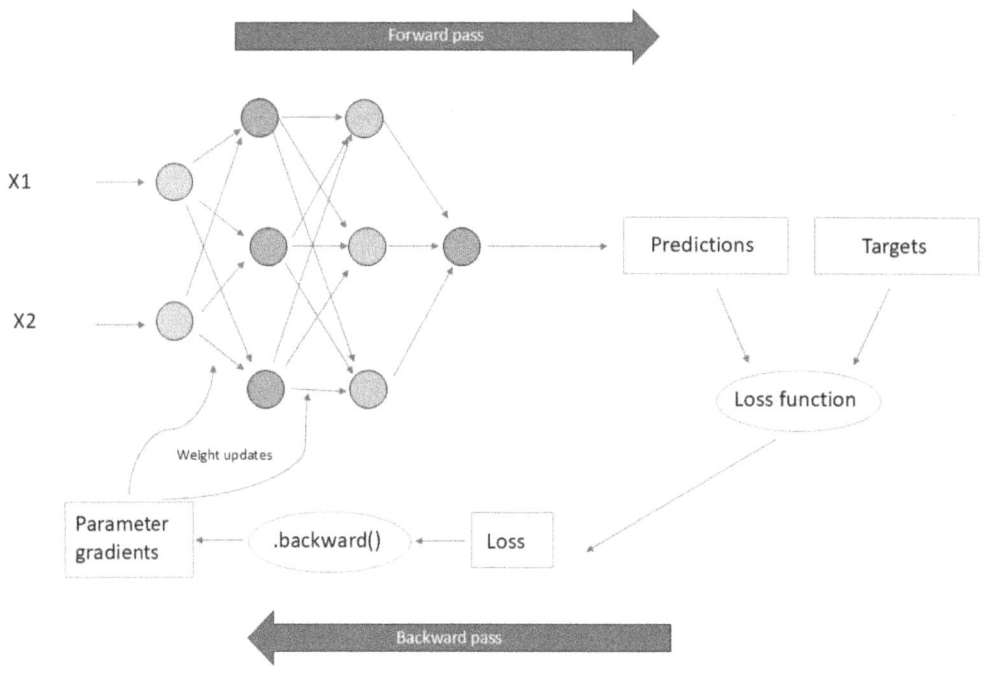

Figure 2-1. *Gradient descent in neural networks*

Exercises in Tensor Manipulations

In this section, we present a few exercises dealing with different types of tensor manipulations. Each exercise corresponds to the use of a single tensor operation or a typical "recipe" which you will use repeatedly in your day-to-day usage of PyTorch. We present these as a question followed by an answer. We recommend that you first try to solve each exercise by yourself and then check the provided solution. PyTorch has a large number of operations, which might be overwhelming and impractical to digest all at once. Going through these exercises meticulously will expose you to a good variety of these operations that are most frequently used in practice.

These exercises are not meant as supplementary material but form the **most important** part of this chapter. Make sure you go over each one of them in order to build a solid intuition of tensors, which will be crucial in understanding the rest of this book.

1. **Define a PyTorch tensor with shape (3,4) containing integers from 1 to 12 sequentially. Get creative to define it as cleanly as possible.**

2. **Change the dtype of the above tensor to float32.**

3. **Transpose its rows and columns (no need to change the input, just return a view).**

4. **Reshape it to take the shape (4,3) (no need to change the input, just return a view).**

Solution:

1. An elegant way to do this is to create a 1D tensor first and then reshape it.

    ```
    a = torch.tensor(range(1,13)).reshape(3,4)
    ```

    Or alternatively:

    ```
    a = torch.arange(1,13).reshape(3,4)
    ```

2. This can be achieved by the to() function.

    ```
    print(f"a.dtype initially: {a.dtype}")
    a = a.to(torch.float32)
    print(f"a.dtype after conversion: {a.dtype}")
    a.dtype initially: torch.int64
    a.dtype after conversion: torch.float32
    ```

    Alternatively, a.float() will achieve the same effect.

3. Multiple ways to achieve this:

    ```
    a.t()
    a.T
    a.transpose(dim0=0,dim1=1)
    tensor([[ 1.,  5.,  9.],
            [ 2.,  6., 10.],
            [ 3.,  7., 11.],
            [ 4.,  8., 12.]])
    ```

Note that this does not change the tensor a but just returns a view with the transpose of a. You will have to set a to one of these to actually modify a.

4. Can be achieved by the reshape function:

```
a.reshape(4,3)
tensor([[ 1,  2,  3],
        [ 4,  5,  6],
        [ 7,  8,  9],
        [10, 11, 12]])
```

Again, this only returns a view of a with shape (4,3): we would have to set a to this in order to change it.

[Masking] Given a PyTorch tensor containing negative and positive integers, replace all the negative integers with -1.
You can work with this tensor:

```
a = torch.tensor([[1, -2, 3], [-4, -3, 0]])
```

Solution:

We first create what is called a Boolean mask. A mask is a Boolean tensor with the same shape as the input, with value True whenever the predicate of interest holds true for an element and False otherwise.

In our case, the mask, which we call neg_mask, would have the value True exactly at those locations which contain negative integers in the input a.

```
neg_mask = a < 0
print(neg_mask)
tensor([[False,  True, False],
        [ True,  True, False]])
```

Verify that it contains True at positions corresponding to values -2, -4, and -3 in a, which are exactly the negative numbers in a. Notice also that the mask is created simply by "a < 0", which compares each element of a to 0 and outputs a Boolean tensor of the same shape as a.

Now that we have located the negative values in a, we replace them with -1 as

```
a[neg_mask] = -1
```

This is a frequently used recipe in PyTorch, where you select elements of the input based on a Boolean mask and set them to a specific value (or perform other operations on them), like input[mask] = new_value.

Print out a and verify that this does the right thing:

```
print(a)
tensor([[ 1, -1,  3],
        [-1, -1,  0]])
```

Given 2D PyTorch tensors a and b, defined as

```
a = torch.tensor([[-3,1,2],[-1,0,6]])
b = torch.tensor([[0,1,0],[1,0,1]])
```

For each of the tensors, do the following:

1. **Check if there is any occurrence of the values 0 or 1.**
2. **Check if all the values of the tensor belong to the set 0, 1.**

Solution:

We first create a tensor containing the elements of interest, i.e., 0 and 1. We then test membership using the isin() function.

```
c = torch.tensor([0,1])
torch.isin(a,c)
tensor([[False,  True, False],
        [False,  True, False]])
```

1. For the first question, we can use the any() function to check if any element in isin() evaluates to True:

   ```
   torch.isin(a,c).any()
   tensor(True)
   ```

   We can use the .item() function to extract the single value "True" from the tensor:

   ```
   torch.isin(a,c).any().item()
   True
   torch.isin(b,c).any().item()
   True
   ```

CHAPTER 2   TENSORS IN PYTORCH

2. We can use the all() function to test if all the elements in the isin() evaluate to True:

    ```
    torch.isin(a,c).all().item()
    False
    torch.isin(b,c).all().item()
    True
    ```

    As expected, all() returns True for b, but False for a.

[Clipping] Given a 2D PyTorch tensor,

1. **Return a tensor with all its elements clipped between 0 and 1. So all the elements less than 0 should be changed to 0, all those greater than 1 should be changed to 1, and those between 0 and 1 should be left unchanged.**
2. **Return a clipping of only its first row between 0 and 1.**
3. **Change the input tensor itself, clipping its first row between 0 and 1.**

You can start with the following input tensor:

```
a = torch.tensor([[-1, 0, 0.2, 1, 1.5],[2, -2, 0, 1, 0.4]])
```

HINT: Use torch.clamp (or torch.clip) function.

Solution:

1. Three equivalent ways of achieving this:

    ```
    print(a.clamp(min=0, max=1))
    print(a.clip(min=0, max=1)) # Alias for clamp()
    print(torch.clamp(a, min=0, max=1))

    tensor([[0.0000, 0.0000, 0.2000, 1.0000, 1.0000],
            [1.0000, 0.0000, 0.0000, 1.0000, 0.4000]])
    tensor([[0.0000, 0.0000, 0.2000, 1.0000, 1.0000],
            [1.0000, 0.0000, 0.0000, 1.0000, 0.4000]])
    tensor([[0.0000, 0.0000, 0.2000, 1.0000, 1.0000],
            [1.0000, 0.0000, 0.0000, 1.0000, 0.4000]])
    ```

CHAPTER 2 TENSORS IN PYTORCH

2. Achieved by simply indexing to choose the first row.

   ```
   print(torch.clamp(a[0,:],min=0, max=1))
   tensor([0.0000, 0.0000, 0.2000, 1.0000, 1.0000])
   ```

3. Using the clamp_() function modifies the input tensor in-place.

   ```
   a[0,:].clamp_(min=0, max=1)
   print(a)
   tensor([[ 0.0000,  0.0000,  0.2000,  1.0000,  1.0000],
           [ 2.0000, -2.0000,  0.0000,  1.0000,  0.4000]])
   ```

Note that most operations have a corresponding function with the same name with the underscore (_) suffix, like torch.mul_, torch.add_, etc. These functions represent the in-place versions of the corresponding operations.

In PyTorch, image datasets are stored in the format (N, C, H, W), where N is the number of images in the set (or the batch), C is the number of channels (usually 3 for a color image, corresponding to red, green, and blue), H is the image height, and W is the image width. Given a tensor containing an image dataset of size [128, 3, 224, 224], perform the following slicing operations:

1. **Select the first 32 images of the set.**

2. **Select only channel 0 for all the samples in the set (note that the output size must be [128, 1, 224, 224], and not [128, 224, 224]).**

3. **For each image in the set, and for all the channels, select pixels with height from 16 to 128 (inclusive of 16, but not 128), and select every alternate pixel from the width, starting from the beginning.**

You can work with a random tensor initialized as follows:

```
image_set = torch.rand([128, 3, 224, 224])
```

Solution:

1) ```image_set[:32, :, :, :].size() # Selects first 32 images```

2) ```image_set[:, 0:1, :, :]```

45

CHAPTER 2   TENSORS IN PYTORCH

Note that if we use indexing instead of slicing, like image_set[:, 0, :, :], it would return a tensor of size [128, 224, 224].

1) image_set[:, :, 16:128, ::2]

Given a 2D tensor, calculate the index of the maximum element in every row. Repeat the same for each column.

HINT: Use torch.argmax function.

You can start with the following input tensor to test your code:

input_tensor = torch.tensor([[1,2,3,4,5,3],[3,5,6,3,2,1]])

Solution:

For the index of the max element along each row:

input_tensor.argmax(dim=1)
>>> tensor([4, 2])

Here, the input to argmax is the dimension, which we specified as 1 for argmax along each row. The output is 4 corresponding to index 4 in 0-based indexing (the fifth element) in the first row, i.e., 5, and 2 corresponding to the second element in the second row, i.e., 6.

To find the index of max elements along each column, we can simply change the dim input to 0 as

input_tensor.argmax(dim=0)
>>> tensor([1, 1, 1, 0, 0, 0])

As expected, the output corresponds to the second row having the maximum element for the first three columns (hence the three 1s), followed by the first row having the maximum element for the later three columns (hence the three 0s).

Given two 1D tensors of the same length, calculate the number of indices where they have the same values.

You can use the following two input tensors to test your code:

target = torch.tensor([1,3,4,5,6,7])
predictions = torch.tensor([2,3,4,6,3,7])

Solution:

The names of the input tensors hint at why this exercise will be of use to us – we will need it to compare how accurate our predictions are compared to the target values.

Let's see what the following simple comparison of two tensors gives:

```
target == predictions
>>> tensor([False, True, True, False, False, True])
```

This gave us a Boolean tensor with True exactly where the two tensors agree. We now just need to add up all the True values. For that, we first convert the above tensor to long:

```
(target == predictions).long()
>>> tensor([0, 1, 1, 0, 0, 1])
```

And then sum it up:

```
(target == predictions).long().sum().item()
>>> 3
```

Note that .sum() returns a 0-dimensional tensor with just one element. Using .item() extracts the value of this tensor.

Given a tensor, return a tensor with all dimensions of size 1 removed.

You can use the following tensor:

```
a = torch.ones([4,1,3,2,1])
```

HINT: torch.squeeze()

Solution:

Using a.squeeze() achieves the result directly. You can verify it by checking their sizes:

```
a.size()
>>> torch.Size([4, 1, 3, 2, 1])
a.squeeze().size()
>>> torch.Size([4, 3, 2])
```

> **Given a 1D tensor of size (n,), return a 2D tensor which has size (n,1), so that each row of this tensor has a single element of the input tensor in the same order.**
> You can work with:
>
> ```
> a = torch.arange(6)
> HINT: torch.unsqueeze()
> ```

Solution:

As the hint suggests, `torch.unsqueeze()` adds a dimension of size 1 at the specified position.

You can verify that this converts the given tensor a

```
tensor([0, 1, 2, 3, 4, 5])
```

into a 2D tensor with each element going to a separate row as

```
a.unsqueeze(dim=1)
tensor([[0],
        [1],
        [2],
        [3],
        [4],
        [5]])
```

Summary

- Tensors are the fundamental data structures in PyTorch. They are essentially multidimensional arrays containing elements of the same data type.

- You can index a tensor to select a "sub-tensor" of lower dimension, which has the dimensions specified in the index as fixed. You can also slice different dimensions of a tensor independently, where slicing in each dimension works similarly to Python lists.

- Tensors are stored in a contiguous block in memory. A tensor is actually a logical view of the contiguous block of memory containing its data.

- Tensors can be initialized in various ways – using lists, NumPy arrays, specialized functions to create tensors filled with ones or zeroes, those filled with random numbers, etc.

- Every PyTorch tensor has three attributes: dtype, device, and layout.

- PyTorch offers a huge collection of tensor operations, including pointwise ops, like mathematical ops, trigonometric functions, logarithmic functions, logical functions, and many others; reduction ops, which reduce the entire tensor to a single scalar value, including several aggregation functions and statistical functions; comparison ops, including those that perform element-wise comparisons, membership operations, sorting, etc.; and several other interesting operations.

- PyTorch's autograd mechanism makes gradient computation very easy and efficient, even for large neural networks containing a chain of operations. PyTorch's autograd, along with out-of-the-box optimizers in torch.optim library, makes the process of training neural networks a breeze for developers.

CHAPTER 3

Image Classification Using Convolutional Neural Networks

This chapter gets the reader acquainted with some essential theoretical concepts on computer vision and convolutional neural networks (CNNs) in particular. It also dives into an image classification project, gradually adding more and more features. The main aim of this chapter is to equip the reader to tackle any new image classification problem confidently.

We start this chapter by describing the image classification problem in computer vision, along with a brief history of the field in the "What Is Image Classification" section. In the "Convolutional Neural Networks (CNNs)" section, we expound on the concepts of convolutional neural networks, introducing the concepts of convolutions, activation functions, padding, pooling, etc. We also explain the concept of data augmentation as a means to inject more variability in the training data to avoid overfitting. In "The Image Classification Project" section, we move on to the practical part of this chapter and bite into our image classification project. We start by loading the dataset and initializing the data loaders, which facilitate iterating over the dataset in batches. We then try to learn more properties of the dataset like the sizes, samples in each class, etc., and also visualize some of the images in the dataset. We then proceed to define our model architectures, iteratively adding complexity to the model. We then write our model training and testing loops and show how to run the end-to-end system. We describe four versions of our model, each being successively more complex than the previous one. We expound on every feature that we add to our model along the way. In the "A More Challenging Dataset" section, we pick a more challenging dataset and run one of our previously defined models on it. Finally, in the "Transfer Learning: Fine-Tuning a Pretrained Model" section, we demonstrate the use of transfer learning by picking a pretrained resnet18 model and fine-tuning the final layer on this challenging dataset.

What Is Image Classification

Image classification is one of the most fundamental problems in computer vision. It involves labeling any given image by one of several predefined class labels. For example, let us say you have a dataset with pictures of either your pet cat or your pet dog, and you want to label each picture as *cat* or *dog*. This problem would fall under the umbrella of image classification.

History

The field of computer vision, in conjunction with trying to understand human vision, has been an active research area since the 1960s. For example, in 1959, Hubel and Wiesel[1] established via experimenting on cats that there are two types of neurons – simple neurons, which process simple structures like oriented edges, and complex ones, which process the higher-level features. Building on this paper, in 1982, David Marr, a British neuroscientist, gave this important insight into the human vision system: he claimed that human vision works in a hierarchical way, that there are low-level algorithms that detect features like edges, curves, etc., and that these are then further processed to detect more and more complex features in a hierarchical way, in order to finally make sense of 3D data[2]. This is a key insight that is used even by the present-day convolutional neural networks (CNNs) on a high level. Inspired again by Hubel and Wiesel's work, a Japanese scientist, Fukushima, designed Neocognitron, the first neural network with convolution layers.

In 1989, Yann Lecun and team introduced the first CNN with backpropagation. Later, in 1998, Lecun and team proposed LeNet5, which was a deep convolutional neural network.

---

[1] Paper: "Receptive fields of single neurons in the cat's striate cortex."
[2] Paper: *Vision: A Computational Investigation into the Human Representation and Processing of Visual Information.*

The next decade saw the advent of much more powerful computing resources, including NVIDIA's GPUs. Another important contribution to the field came from large, labelled datasets, notably ImageNet, with 1000 categories and a few million images, introduced by Fei-Fei Li in 2009.

Since 2010, ImageNet has been used for the ImageNet Large Scale Visual Recognition Challenge (ILSVRC), which became a benchmark to test state-of-the-art image classification and object detection models. The year 2012 saw the resurgence of CNNs when AlexNet achieved a remarkably better performance than the prevalent models of the time (which were based on classical techniques like SVMs, etc.). Since then, this field has been dominated by CNNs, which have evolved over time, and there have been many improved versions like ResNets (2016), MobileNet (2017), etc.

Convolutional Neural Networks (CNNs)

Convolutional neural networks are a type of deep neural network architecture prominently used in computer vision tasks. We shall study the main components of these architectures in this section.

Convolutions

Convolution is a mathematical operation applied to an image (or to intermediate feature maps, as we will see later) in a neural network. It involves what are called **filters or kernels**, which are small matrices (3 × 3, 5 × 5, etc.) containing **weights**. A filter slides over the image, aligning with each submatrix of the image of the same size as the filter. At each such position, it takes the element-wise product of the filter weights and the corresponding pixel values and adds them together. Each result is written in an output matrix in the same position.

This will become clearer with an example:

Let's say we have a 7×7 grayscale image[3] and a 3 × 3 filter as depicted in Figure 3-1.

---

[3] These pixel numbers are arbitrarily chosen and don't actually correspond to any image.

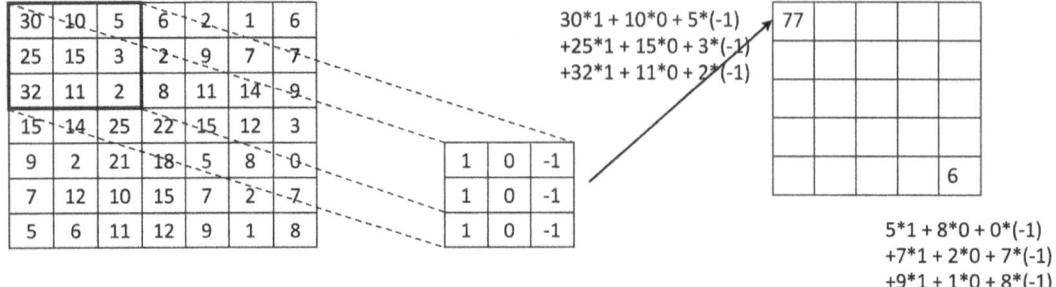

Figure 3-1. *Convolution*

Here, consider the 3×3 submatrix in the top left corner: as the filter aligns with this submatrix, the filter weight multiplies with the corresponding pixel value, and all these products are added together as shown in the picture. This gives 30*1 + 10*0 + 5*(-1) + ... = 77, which is then filled in in the top left corner of the output matrix.

The filter then moves over by one position to align with the submatrix in Figure 3-2.

10	5	6
15	3	2
11	2	8

Figure 3-2. *Submatrix*

The result of the products of this with the weights, followed by their sum, is then stored in the next position and so on.

Vertical and Horizontal Edge Detection

In the above example, the kernel we used detects vertical edges. For example, observe that the top left 3×3 submatrix contains very high values in the left corresponding to white color and very low values in the right corresponding to black. This sharp change of color is detected by this filter, as it gives a very high value of 77 in the corresponding position. On the contrary, if you see the bottom right 3×3 corner of the image, it has pixels of similar values, indicating that there is no edge present there. Our kernel outputs a low value of 6 in that position, thereby not indicating any edge as expected. Applying a simple threshold activation function on top of this will output a 1 where a vertical edge is present and a 0 elsewhere (actually, we would need to also apply absolute value first to detect vertical edges where the left part is dark and the right part is bright).

CHAPTER 3 IMAGE CLASSIFICATION USING CONVOLUTIONAL NEURAL NETWORKS

Analogously, horizontal edges can be detected by a filter as shown in Figure 3-3.

1	0	-1
1	0	-1
1	0	-1

Vertical edge detector

1	1	1
0	0	0
-1	-1	-1

Horizontal edge detector

Figure 3-3. *Horizontal and vertical edge filters*

Note that there are many other filters too that will detect horizontal or vertical edges; these were just picked as representatives.

Activation Functions

After convolving in this way, an activation function is applied to each output value in this matrix to introduce nonlinearity. Usually, we will be using the ReLU (rectified linear unit) activation function in our architectures. Each such output matrix is called a *feature map* since it encodes certain features of the input image depending on the filter chosen. Alternatively, it is also sometimes referred to as an activation map.

In fact, in each layer, we can have multiple filters that detect different types of features. For example, we can have one filter for vertical edges, one for horizontal edges, one for curvatures, etc. Each filter would output a feature map matrix. The collection of such feature map output matrices is stacked together as a *cube* or a *volume.*

Each such collection of convolution operations, followed by an activation function, is called a **convolution layer**.

We apply a convolution layer not just to the input image but also to intermediate feature maps. Remember that the 2D feature map matrices we saw before are now stacked together to give 3D volumes of feature maps. Therefore, our filters would also have to be 3D with depth matching that of the input feature map on which it is applied. The functioning of the filter is still the same – it multiplies each filter-weight by the corresponding pixel value element-wise and adds them all together. Each filter therefore still outputs a 2D matrix. The depth of the output feature map is equal to the number of filters used in each layer.

Note that if the input image is n × n and we slide a 3 × 3 filter over it, the output matrix will be n - 2 × n - 2. More generally, if the input feature map is n × n × d and we slide an f × f filter over it, the output 2D feature map has size (n – f + 1) × (n – f + 1).

55

CHAPTER 3 IMAGE CLASSIFICATION USING CONVOLUTIONAL NEURAL NETWORKS

Strides

We also have the option of sliding the filter multiple pixels horizontally and vertically. The number of pixels by which we move the filter at each step is called the *stride*. In our examples so far, we have used a stride of 1, sliding the filter one pixel at a time in each direction. We could also move it two pixels at a time, effectively "skipping" positions and producing a smaller output — this corresponds to a stride of 2. In this book, we will always use a stride of 1 for simplicity.

Padding

Each convolution layer shrinks the width and height of the image (or feature map), as per the calculations we just saw. To avoid this, we can pad the image along the border by some fixed value, typically 0. For example, let us say we are using a 3 × 3 filter with a stride of 1. If we pad the input image by one row below and above the image and one column to the left and right of the image, the effective dimension of the image on which we apply the convolution would be (n+2) × (n+2). Therefore, after convolution, its dimension would be (n + 2 - 2) × (n + 2 - 2) = n × n, maintaining the original height and width. In general, padding by p rows and columns on each boundary followed by convolution by an (f × f) filter gives an output of dimension (n + 2p - f +1) × (n + 2p - f +1). Therefore, to keep the height and width of the input unchanged, we can apply a padding of p = (f - 1)/2. Such a convolution, which maintains the width and height of the input, is called a *same* convolution.

More generally, the output size also depends on the stride. For a 2D input of size n × n, a filter of size f × f, padding p per boundary, and stride s, the output feature map has spatial dimension given by $\left| \frac{(n - f + 2p)}{s} \right| + 1$.

For example, for an input of size 28 × 28, with filter size 3, padding 1, and stride 1, the output size remains 28 × 28. With stride 2, the output shrinks to 14 × 14.

Dimension Calculations

Let us say that at layer i in the network, we have feature maps of dimension $h_i \times w_i \times d_i$. If we use the *same* convolution, the output would also have the same width and height. The depth of the output feature map would match the number of filters used in the current layer.

Also, the depth of each current filter would have to match d_i. If we use a filter of dimension $f_i \times f_i \times d_i$, then we would have to use padding of size $p = \frac{f_i - 1}{2}$ for same convolution, assuming a stride of 1.

All these calculations will prove to be quite important later when we construct a CNN architecture with multiple convolution layers.

These filter weights are treated as model parameters, which are learned via backpropagation using the gradient descent algorithm.

Pooling

After one or more convolution layers, a pooling layer is usually added to reduce the dimension of the feature map. It also helps in regularization and helps in creating a hierarchical feature representation. A p × p pooling layer partitions each channel of the input feature maps into p × p matrices and applies an aggregate function on each of these p × p matrices to get a $\lceil n/p \rceil \times \lceil n/p \rceil$ feature map.

Let us take the example of a 2×2 max pooling layer. Each channel of the feature map is partitioned into 2×2 nonoverlapping boxes. The max pooling layer then takes the max value of each of these 2×2 boxes and selects it in its output feature map, as illustrated in Figure 3-4.

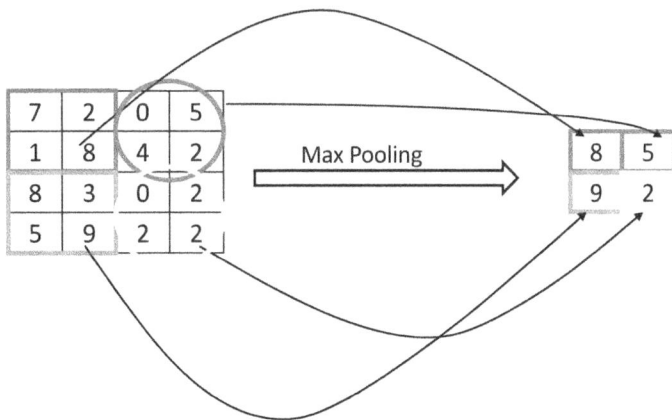

Figure 3-4. *Max pooling example*

This reduces the height and width by half while keeping the number of channels unchanged.

CHAPTER 3   IMAGE CLASSIFICATION USING CONVOLUTIONAL NEURAL NETWORKS

2×2 max pooling is one of the most commonly used pooling layers. One can also apply pooling of other dimensions. Another common variation of the aggregate function is the average pooling layer, which applies the average function instead of taking the max.

Note that the pooling layer does not have any learnable parameters.

Flatten

After applying a few convolution and pooling layers, we come to the final parts of our neural network, where we first apply a Flatten layer. By this point, we have usually reduced the width and height by a significant factor by applying pooling layers and have increased the number of channels. The Flatten layer, as the name suggests, simply flattens out the entire volume of the feature map into a long 1D vector.

Fully Connected or Dense Layer

After the Flatten layer, we usually use a fully connected (i.e., dense) layer. A dense layer is one where every input is connected to every output of the layer. Each neuron computes a linear combination of the inputs, followed by a nonlinearity such as ReLU.

This layer takes the dimension of the Flatten layer as input, whereas its output is the number of classes in the image classification task. For example, the handwritten digits task, which we will work with, will contain ten outputs corresponding to the ten labels $\{0,1,...9\}$.

Softmax

The fully connected layer outputs ten scores if we have ten labels. We typically interpret this as the higher the score, the greater the chance of that being the true label. However, these scores can be any real numbers. How can we predict these as probabilities?

We use what is called the softmax function for this. The first step here is to convert each of the scores $(s_0, s_1, ...s_9)$ into exponentials: $(e^{s_0}, e^{s_1}, ...e^{s_9})$. These exponentials are all positive and can be thought of as unnormalized probabilities. We then normalize these to get probabilities as $(\frac{e^{s_0}}{\sum_{0}^{9} e^{s_i}}, \frac{e^{s_1}}{\sum_{0}^{9} e^{s_i}}, ...\frac{e^{s_9}}{\sum_{0}^{9} e^{s_i}})$.

For example, the first score now gives the probability of the label being 0 as predicted by the model, and so on.

Cross-Entropy Loss

We shall use the cross-entropy loss or log-loss function in our CNN.

Our model, after applying softmax, predicts a probability distribution over class labels. The target label values can also be thought of as a probability distribution, which has a probability of 1 for the actual label for an input image and a probability of 0 for all other labels.

The cross-entropy loss is a way of measuring the dissimilarity between these two distributions. Mathematically, if p_i is the predicted probability of each class label i and y_i is the distribution described above for the ground truth (i.e., $y_i = 1$ for an image with target label i, and $y_i = 0$ otherwise).

Then the cross-entropy loss, if we have k classes, is defined as

$$CrossEntropy(y,p) = -\sum_{i=1}^{k} y_i \log p_i$$

For our case, since y_i is 1 only for the class label of an image, this would just be $-\log p_i$ where i is the label.

Batch Normalization Layer

A batch normalization layer[4] or *batchnorm* is applied after a convolution block and before the nonlinear activation. It involves normalizing the data in the layer for each batch. It is a means of addressing the variation in distributions of inputs to a layer across different batches, which is called the internal covariate shift.

More specific to the context of CNNs, for each channel, it computes the mean and standard deviation for the entire batch across all the values of that channel and normalizes the values in that channel according to this mean and standard deviation. Normalizing essentially subtracts the mean from each value and divides it by the standard deviation, ensuring that the resultant values have a mean of zero and standard deviation of 1. After normalization, these values are then scaled using a learnable parameter gamma and shifted by a learnable parameter beta. These parameters, gamma and beta, are learned during the training process.

---

[4] The paper which introduced batch normalization: "**Batch Normalization: Accelerating Deep Network Training by Reducing Internal Covariate Shift.**"

Batch normalization offers several benefits like

1. Allows using higher learning rates
2. Allows being less careful about initialization
3. Speeds up the training process
4. Acts as a regularizer (helps avoid overfitting)

It is worth noting that while the batch normalization layer uses the mean and standard deviation of each batch for normalization during training, it operates slightly differently during the inference (or testing) phase. During training, the model keeps track of moving averages of the mean and standard deviations of the activations in each batchnorm layer. During testing (and inference), it uses these moving averages of the mean and standard deviation for normalization, instead of using the mini-batch statistics like it does in training.

Dropout Layer

Dropout[5] is a way to regularize the model, i.e., avoid overfitting. A dropout layer is usually applied after the nonlinear activation. It is applied only during training and **turned off during inference**. It is characterized by a hyperparameter p, and during each forward pass, it randomly sets to 0 each element of the activation with probability p. Intuitively, this prevents the network from relying too heavily on a few select neurons for its output and helps to avoid overfitting.

Data Augmentation

Sometimes, the number of samples in the training dataset is not sufficient to learn the patterns of the underlying population. In such cases, we may perturb the available training data in different ways so that the model sees different variations of the same data samples, which helps it become more generalizable. This process is called data augmentation. Since this exposes the model to more variants of the data, it helps in avoiding overfitting.

---

[5] Paper which introduced dropout: **"Dropout: A Simple Way to Prevent Neural Networks from Overfitting."**

For image datasets, several different data augmentation techniques are used in practice as follows:

1. Geometric transforms include

    a. Flipping images horizontally or vertically at random

    b. Rotating images at random angles

    c. Translation: Moving images around

    d. Resizing: Randomly cropping or resizing images

2. Color transforms: These include randomly changing the brightness, contrast, hue, and saturation, randomly permuting color channels, converting images to grayscale, etc.

3. Noise addition: Adding Gaussian (or other types of) noise to images.

The Image Classification Project

Dataset

PyTorch offers a library of datasets for computer vision in the torchvision.datasets module. Apart from that, it also offers functionalities that make it easy for you to use your own dataset. For our project, we will start simple and choose the MNIST dataset, which PyTorch provides. This dataset comprises grayscale images of handwritten digits. In the latter part of the chapter, we will work with the more complex Oxford-IIIT pet dataset.

Main Components of the Project

This project can be broken down into four main components:

1. Loading the data into DataLoaders

2. Data exploration and visualization

3. Defining the model architecture

4. Model training, validation, and testing loops

We shall build each component separately and then put them all together to see how it works. We shall look at four versions of the models, each version being incrementally more complex than the last.

Imports

We first list all the imports that you will need for this project:

```
import torch
from torch.utils.data import Dataset
from torchvision.transforms import ToTensor
import matplotlib.pyplot as plt
from torch import nn
from torchvision import datasets
from torch.utils.data import DataLoader
```

Loading Data from Torch Datasets

We first load the MNIST dataset here.

```
training_data = datasets.MNIST(
    root="data",
    train=True,
    download=True,
    transform=ToTensor()
)

test_data = datasets.MNIST(
    root="data",
    train=False,
    download=True,
    transform=ToTensor()
)
```

Here, the argument *root* is where the dataset will be saved on your machine if *download* is set to True, which is the case for us. Setting *train* = True loads the training dataset, whereas train = False loads the test component of the dataset. It also offers a transform argument, which allows you to transform the input images before loading them. We just use the ToTensor() transform here, which converts the images to PyTorch tensors.

Printing these out gives some information about the datasets, the transforms we used, etc.:

```
print(train_data)
Dataset MNIST
    Number of datapoints: 60000
    Root location: data
    Split: Train
    StandardTransform
Transform: ToTensor()

print(test_data)
Dataset MNIST
    Number of datapoints: 10000
    Root location: data
    Split: Test
    StandardTransform
Transform: ToTensor()
```

DataLoaders

A dataset object, as we saw before, stores all the transformed data samples. PyTorch also offers a neat way of iterating over datasets by what are called data loaders.

We will now wrap the datasets we loaded into these iterable DataLoaders, which will enable us to iterate batch-wise through the entire dataset.

```
train_dataloader = DataLoader(training_data, batch_size=64, shuffle=True)
test_dataloader = DataLoader(test_data, batch_size=64, shuffle=False)
```

As you can see, it also offers an option to set the batch_size, which will help us to pass the data batch-wise through the model while using the mini-batch gradient descent algorithm. It also has a shuffle argument, which, when set to True, shuffles the data after every epoch (recall that an epoch corresponds to one pass of the training data through the model).

CHAPTER 3   IMAGE CLASSIFICATION USING CONVOLUTIONAL NEURAL NETWORKS

```
train_features, train_labels = next(iter(train_dataloader))
print(f"Training features batch shape: {train_features.size()}")
print(f"Training Labels batch shape: {train_labels.size()}")

>>> Training features batch shape: torch.Size([64, 1, 28, 28])
>>> Training Labels batch shape: torch.Size([64])
```

Using *iter* on a DataLoader creates an iterable, and you can access the next batch of 64 samples using the *next* function on the iterable. Train_features tensor is of size [64, 1, 28, 28], corresponding to 64 samples in the batch, one channel since it contains grayscale images, and 28 × 28 being the size (height × width) of each image. The labels of the training set simply contain 64 scalar values.

Data Exploration and Visualization

We first plot a sample image from each label, going from 0 to 9, in Figure 3-5. This will give us a sense of the images that we will work with.

```
figure = plt.figure(figsize=(5, 2.5))
for i in range(10):
    for j in range(len(training_data)):
        img, label = training_data[j]
        if label == i:
            figure.add_subplot(2, 5, i+1)
            plt.title(label)
            plt.axis("off")
            plt.imshow(img.squeeze(), cmap="gray")
            break
plt.show()
```

CHAPTER 3 IMAGE CLASSIFICATION USING CONVOLUTIONAL NEURAL NETWORKS

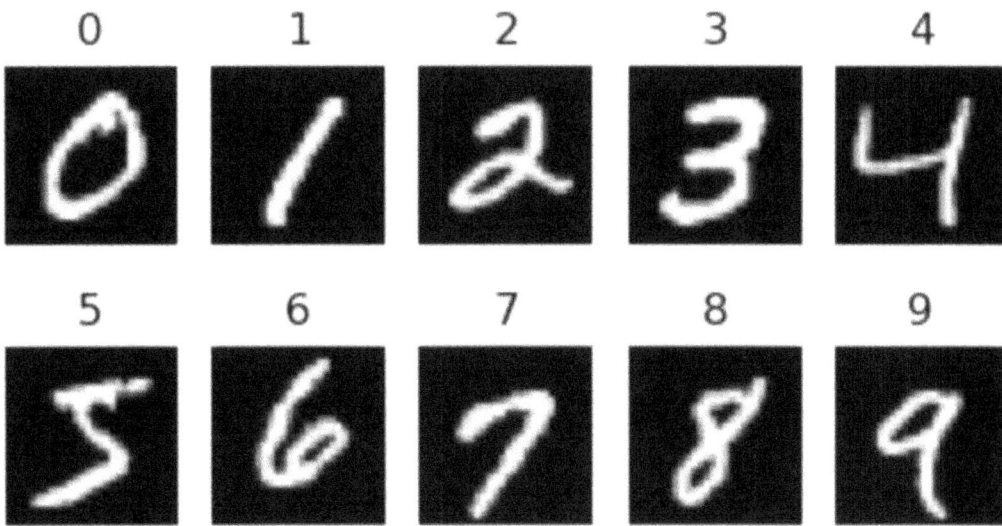

Figure 3-5. *Example images in the MNIST dataset*

We also plot a histogram of the label frequencies to check if there is any class imbalance (see Figure 3-6).

```
# Create a PyTorch tensor of labels
labels_list = [label for features, label in training_data]
labels_tensor = torch.tensor(labels_list)

# Count the frequency of each label
unique_labels, counts = torch.unique(labels_tensor, return_counts=True)

# Plotting the histogram
plt.bar(unique_labels, counts,tick_label=list(range(10)))
plt.xlabel("Label")
plt.ylabel("Count")
plt.title("Label Distribution in training data")

plt.show()
```

CHAPTER 3   IMAGE CLASSIFICATION USING CONVOLUTIONAL NEURAL NETWORKS

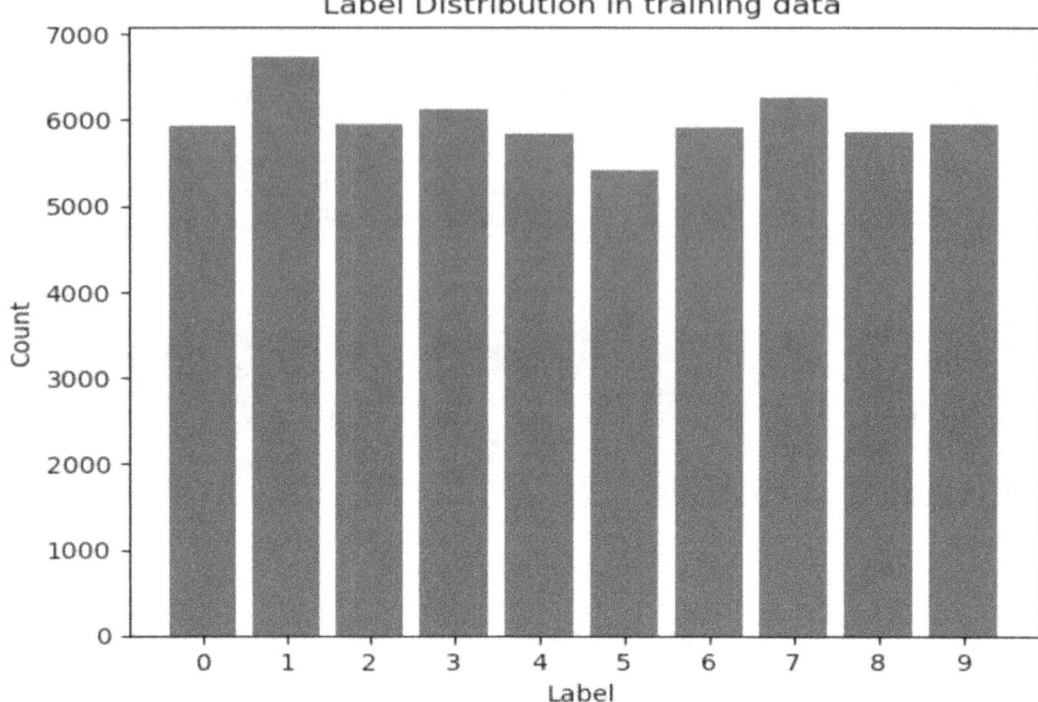

Figure 3-6. *Count of each label in the MNIST dataset*

Each label has about 5000–6000 images, without a significant class imbalance.

Model Definition

We now come to the most creative part of the project, which is model architecture design. We first begin with a simple linear classification model architecture, which just flattens the input image into a 28*28-sized tensor and then simply uses a single dense (i.e., fully connected) layer with ten outputs. It then uses the cross-entropy loss for training. We don't need to explicitly apply the softmax function to the output, since PyTorch's CrossEntropyLoss implicitly applies a log_softmax().

```
class ConvNeuralNetwork1(nn.Module):
    def __init__(self):
        super().__init__()
        
        self.flatten = nn.Flatten() # Flattens all but dim 0
        self.fully_connected = nn.Linear(28*28, 10)
```

```
def forward(self, x):
    # Input x has size [64, 1, 28, 28]
    # Each batch has 64 images of size 28x28, with 1 channel each.
    x = self.flatten(x) # x.size() >>> torch.Size([64, 784])
    x = self.fully_connected(x) # x.size() >>> torch.Size([64, 10])

    return x
```

We now use the above simple linear model[6] as a running example to explain some important considerations to keep in mind while defining any neural network architecture in PyTorch:

1. Your model should be a subclass of nn.Module.

2. Your model should contain these two functions: __init__() and forward().

3. __init__: Is used to initialize all the layers of your model, along with the parameters, sizes, etc. It is important to begin the function definition with a call to super().__init__: this calls the init function of the base class (i.e., nn.Module), which initializes the attributes, etc., essential for your model training and inference.

4. You can then define the different layers of your neural network as attributes. In this case, we only have self.flatten and self.fully_connected. You don't need to implement each of these layers by hand, since PyTorch provides all these functionalities (and many more) as part of the torch.nn package.

5. Note once again that we don't need to explicitly add a softmax layer, since PyTorch's CrossEntropyLoss applies a log_softmax() internally.

6. nn.Flatten by default flattens along all dimensions but dimension 0 - this is because dimension 0 corresponds to different samples in our batch, and so it wouldn't make sense to flatten along this dimension.

---

[6] Note that we call this model ConvNeuralNetwork1, although technically there is no convolution involved in this simplest version of the model.

CHAPTER 3 IMAGE CLASSIFICATION USING CONVOLUTIONAL NEURAL NETWORKS

7. While initializing nn.fully_connected, which simply contains a linear layer with 28*28 inputs and ten outputs.

8. forward(): The forward() function is called whenever the model is called during inference or during the forward pass while training. It takes the input features x as its arguments. The model's computation graph is defined by the operations within this function, and PyTorch's autograd mechanism calculates gradients during the backward pass based on operations defined in this forward function.

9. While implementing forward, you can just pass the input x to each of your model layers defined during the init function sequentially.

Model Training Loop

We now define the model training loop via a function called train_loop.

```
def train_loop(dataloader, model, loss_fn, optimizer):
    data_size = len(dataloader.dataset)
    batch_size = dataloader.batch_size
    model.train() # Doesn't train the model, but sets it to train mode for
    BatchNorm layers
    total_loss = 0
    print(f"Training set:")
    for batch_idx, batch_data in enumerate(dataloader):
        # Compute predictions and loss for this batch
        X, y = batch_data
        pred = model(X)
        loss = loss_fn(pred, y)
        total_loss += loss.item()

        # Backpropagation and parameter updates
        loss.backward()
        optimizer.step()
        optimizer.zero_grad()

        # Print loss for every 100 batches
```

```
    if batch_idx % 100 == 99:
        print(f"Loss for batch {batch_idx+1:>3d}: {total_
        loss/100:>4f}")
        total_loss = 0
```

    This function takes as input the dataloader, the model, a loss function, and an optimizer. Each call to the training loop passes the entire training data once through the model. A call to the training loop thus corresponds to one epoch. Using the dataloader passed to it, it goes through the entire data one batch at a time. It passes each batch of 64 samples (since we specified the batch_size to be 64 to the data loader) through the model to get the model predictions with the current parameter settings. It then computes the loss using the loss function. Finally, it calls loss.backward() to initiate backpropagation, and the optimizer updates the parameters using the gradients populated in .grad attributes during the call to backward() function. Optimizer.zero_grad() then resets all the parameter gradients to zero once the parameters have been updated. This is necessary because, after each call to backward(), PyTorch accumulates the gradients by adding them to the current gradient values. Finally, the training loss is printed for every 100 batches.

    Note that before the loop, there is a call to model.train(). This might be misleading because this does not actually train the model in any way but just sets the model to training mode, which will be relevant when we use BatchNorm layers later.

Test Loop

After each epoch, we also print the loss and accuracy on the test set using the test_loop:

```
def test_loop(dataloader, model, loss_fn):
    data_size = len(dataloader.dataset)
    batch_size = dataloader.batch_size
    model.eval() # Sets the model to eval (non-training) mode for
    BatchNorm layers

    num_batches = len(dataloader)
    test_loss, correct_pred = 0, 0

    with torch.no_grad(): # Don't need gradients during inference
        for X, y in dataloader:
            pred_logits = model(X)
```

```
            test_loss += loss_fn(pred_logits, y).item()
            pred_label = pred_logits.argmax(1)
            correct_pred += (pred_label == y).long().sum().item() # Refer
            to exercise in chapter 2
    test_loss /= num_batches
    pred_accuracy = 100.0*correct_pred / data_size
    print(f"Test set: \n Accuracy: {pred_accuracy:>0.1f}%, Avg loss:
    {test_loss:>6f} \n")
```

The model.eval() line sets the model to evaluation mode or non-training mode. This is important for batchnorm layers, which we will see later. Also, we pass the data through our model inside a torch.no_grad() declaration – this tells PyTorch to not take the gradients during all these tensor operations, thereby saving time and compute memory.

Putting It All Together

Now that we have put together the engine of our car, it is time to make the wheels turn! We first define the hyperparameters like the learning rate, batch_size, and number of epochs. We also fix torch.manual_seed to reduce the randomness of each run. We now call the train_loop function followed by the test_loop function in a for loop, which is run for num_epochs iterations.

```
torch.manual_seed(9) # To reduce variability in each run
# Hyperparameters
learning_rate = 0.05
batch_size = 64
num_epochs = 40

model = ConvNeuralNetwork1()
loss_fn = nn.CrossEntropyLoss()
optimizer = torch.optim.SGD(model.parameters(), lr=learning_rate)

for i in range(num_epochs):
    print(f"Epoch {i+1}\n---------------------------------------------")
    train_loop(train_dataloader, model, loss_fn, optimizer)
    test_loop(test_dataloader, model, loss_fn)
print("Model trained!")
```

Results

At the end of 40 epochs, this model gave me an accuracy of about 92.3% on the test set. This already seems pretty encouraging, given that all we used was a dense layer, with no convolutions so far. To get a sense of how good this is, note that a random prediction of the label from 0 to 9 would give an accuracy of only about 10% for a perfectly balanced set.

Refining the Model
Model 2: Using a Convolution Layer

We now add a convolution and a pooling layer to our previous model:

```
class ConvNeuralNetwork2(nn.Module):
    def __init__(self):
        super().__init__()
        self.conv1 = nn.Conv2d(in_channels=1, out_channels=64, kernel_size=(3,3), padding=1)
        self.relu_act = nn.ReLU()
        self.max_pool = nn.MaxPool2d(kernel_size=(2,2))
        self.flatten = nn.Flatten()
        self.fully_connected = nn.Linear(14*14*64, 10)

    def forward(self, x):
        x = self.conv1(x)
        x = self.relu_act(x)
        x = self.max_pool(x)
        x = self.flatten(x)
        x = self.fully_connected(x)
        return x
```

CHAPTER 3   IMAGE CLASSIFICATION USING CONVOLUTIONAL NEURAL NETWORKS

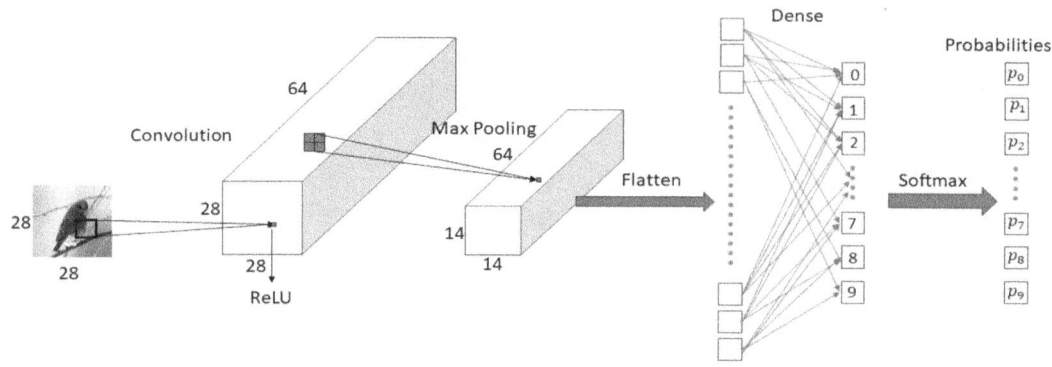

Figure 3-7. *Model architecture for ConvNeuralNetwork2*

The first layer applies a 3 × 3 convolution with padding of one, which maintains the width and height to 28 × 28. It has 64 output channels, so the result of this is a 28 × 28 × 64-sized tensor (for each data sample).

Then the ReLU activation function is applied. We then apply a max pooling layer with a 2×2 kernel, which halves the width and height to 14 × 14, with the number of channels still being 64. We then use the Flatten layer, which converts it into a 1D tensor (per sample) of size 14*14*64 = 12544. I would recommend that you always try to keep these calculations (like the multiplication) as is in the code for readability.

We then apply a dense or fully connected layer with ten outputs. Finally, we use softmax (implicitly applied by CrossEntropyLoss in PyTorch) to convert these scores into probabilities. For calculating the loss, we shall be considering the argmax of the output as the predicted label. In other words, the label with the highest predicted probability will be considered as the model's prediction. See Figure 3-7 for an illustration of the model architecture.

Running this on my machine gave an accuracy of a whopping 98.4% on the test data.

Model 3: Using Three Convolution Layers

We now make our model architecture more complex by adding three consecutive units of convolution-RELU-maxpool, followed, as before, by the Flatten and dense layers.

```
class ConvNeuralNetwork3(nn.Module):
    def __init__(self):
        super().__init__()
        # First convolution-block
```

```python
        self.conv1 = nn.Conv2d(in_channels=1, out_channels=32, kernel_
        size=(3,3), padding=1)
        self.relu_act1 = nn.ReLU()
        self.max_pool1 = nn.MaxPool2d(kernel_size=(2,2))

        # Second convolution-block
        self.conv2 = nn.Conv2d(in_channels=32, out_channels=64, kernel_
        size=(3,3), padding=1)
        self.relu_act2 = nn.ReLU()
        self.max_pool2 = nn.MaxPool2d(kernel_size=(2,2))

        # Third convolution-block
        self.conv3 = nn.Conv2d(in_channels=64, out_channels=128, kernel_
        size=(5,5), padding=2)
        self.relu_act3 = nn.ReLU()
        self.max_pool3 = nn.MaxPool2d(kernel_size=(2,2))

        self.flatten = nn.Flatten()
        self.fully_connected = nn.Linear(3*3*128, 10)

    def forward(self, x):
        x = self.conv1(x)
        x = self.relu_act1(x)
        x = self.max_pool1(x)

        x = self.conv2(x)
        x = self.relu_act2(x)
        x = self.max_pool2(x)

        x = self.conv3(x)
        x = self.relu_act3(x)
        x = self.max_pool3(x)

        x = self.flatten(x)
        x = self.fully_connected(x)
        return x
```

Observe that since we use the same convolutions, the width and height of feature maps are left unchanged by the convolution layers. Each map pool layer halves the width as well as height, so the three map pool layers result in feature maps with height (and width) of $\lfloor\lfloor\lfloor 28/2\rfloor/2\rfloor/2\rfloor = 3$.

After 40 epochs, model 3 yields an accuracy of 99.0%.

Further Improvements to the Model and the Training Process

We now add two improvements to the training process, i.e., early stopping and learning rate scheduling, and two more improvements to our model, i.e., batch normalization layer and dropout layer. These would help us to stabilize and speed up the learning process, avoid overfitting, and help in model convergence.

Early Stopping

Early stopping is based on the principle that the model should only train so far as there is some improvement in the validation metrics. Therefore, we begin by further splitting the training data into training and validation data (9:1 ratio).

```
train_validation_ratio = 0.1

train_data, val_data = random_split(train_data, [1 - train_validation_ratio, train_validation_ratio])

train_dataloader = DataLoader(train_data, batch_size=64, shuffle=True)
val_dataloader = DataLoader(val_data, batch_size=64, shuffle=False)

test_dataloader = DataLoader(test_data, batch_size=64, shuffle=False)
```

We then define an EarlyStopper class.

```
class EarlyStopper:
    def __init__(self, min_delta = 0.001, patience = 5):
        self.patience = patience
        self.min_delta = min_delta
        self.improvement_counter = 0
        self.best_val_loss = float('inf')
```

```
def should_stop(self, val_loss):
    if val_loss < self.best_val_loss - self.min_delta:
        # sufficient improvement
        self.improvement_counter = 0
        self.best_val_loss = val_loss
    else:
        self.improvement_counter += 1
        if self.improvement_counter >= self.patience:
            return True
    return False
```

Our EarlyStopper class has four attributes:

1. The patience parameter (defaults to 5). This signifies the number of epochs for which we are willing to tolerate a degradation of validation loss (or more precisely, an insufficient improvement in validation loss).

2. Patience is implemented by an improvement_counter, which is set to 0 every time we see an improvement in validation loss, and it goes on increasing every time we don't see a significant improvement.

3. The best validation loss so far is stored in a variable called best_val_loss.

4. The parameter min_delta, defaulting to 0.001, helps inform what we should consider as a *significant* improvement in validation loss – the validation loss needs to be at least min_delta lower than the best loss so far to count as significant.

The should_stop function takes the validation loss as input and checks if the validation loss has improved over the best validation loss so far by more than min_delta. If it hasn't improved for more than *patience* consecutive epochs, it returns True, and otherwise, it returns False.

CHAPTER 3 IMAGE CLASSIFICATION USING CONVOLUTIONAL NEURAL NETWORKS

We also define a validation loop on similar lines as the test loop, which ends up calling the should_stop function of our EarlyStopper:

```
def val_loop(dataloader, model, loss_fn, early_stopper):
    data_size = len(dataloader.dataset)
    batch_size = dataloader.batch_size
    model.eval() # Sets the model to eval (non-training) mode for BatchNorm layers

    num_batches = len(dataloader)
    val_loss, correct_pred = 0, 0

    with torch.no_grad(): # Don't need gradients during validation
        for X, y in dataloader:
            pred_logits = model(X) # prediction probabilities
            val_loss += loss_fn(pred_logits, y).item()
            pred_label = pred_logits.argmax(1)
            correct_pred += (pred_label == y).long().sum().item() # Refer to exercise in chapter 2
    val_loss /= num_batches
    pred_accuracy = 100.0*correct_pred / data_size
    print(f"Validation set: \n Accuracy: {pred_accuracy:>0.1f}%, Avg loss: {val_loss:>6f} \n")
    if early_stopper.should_stop(val_loss):
        return True
    return False
```

We also need to modify the code that calls the trainer, etc., as below:

```
torch.manual_seed(9) # To reduce variability in each run
# Hyperparameters
learning_rate = 0.05
batch_size = 64
num_epochs = 40

model = ConvNeuralNetwork3()
loss_fn = nn.CrossEntropyLoss()
optimizer = torch.optim.SGD(model.parameters(), lr=learning_rate)
early_stopper = EarlyStopper()
```

```
for i in range(num_epochs):
    print(f"Epoch {i+1}\n-------------------------------------------")
    train_loop(train_dataloader, model, loss_fn, optimizer)
    test_loop(test_dataloader, model, loss_fn)
    if val_loop(val_dataloader, model, loss_fn, early_stopper):
        print(f"Early stopping.")
        break
print("Model trained!")
```

To summarize this process, the should_stop() method of our early_stopper is called on the validation data after every epoch, and it returns True if for the last five (patience) epochs, the validation loss hasn't improved over the best validation loss so far by more than 0.001 (min_delta). In case should_stop returns true, the model training is stopped after that epoch.

Note that it is common practice to choose the model with the lowest validation loss so far after early stopping, but we skip that step here to avoid clutter.

Early stopping potentially reduces the training time, but more importantly, it ensures that the model does not overfit on the training data (refer to overfitting in Chapter 1).

Learning Rate Scheduler

The success of the optimization (gradient descent) algorithm depends heavily on the learning rate. If the learning rate is too high, the gradient descent algorithm will take large jumps around the minima, and it is unlikely to ever converge to the minima. On the other hand, if the learning rate is too small, the gradient descent algorithm may take tiny steps, taking forever to converge. Therefore, it is often useful to start with a high enough learning rate and decay it as the algorithm proceeds. This process is called learning rate scheduling. Learning rate scheduling helps model convergence, helps improve the model accuracy, and also helps reduce overfitting.

PyTorch offers several options for learning rate scheduling in its torch.optim module. Here, we use the exponential learning rate scheduler, which decays the learning rate after every epoch by the multiplicative factor of gamma, which it takes as an argument. Here, we set gamma to 0.9. The scheduler.step() line called at the end of each epoch updates the learning rate.

```
from torch.optim.lr_scheduler import ExponentialLR
torch.manual_seed(9) # To reduce variability in each run

# Hyperparameters
learning_rate = 0.05
batch_size = 64
num_epochs = 40

#Defining model, loss, optimizer
model = ConvNeuralNetwork4()
loss_fn = nn.CrossEntropyLoss()
optimizer = torch.optim.SGD(model.parameters(), lr=learning_rate)
scheduler = ExponentialLR(optimizer, gamma=0.9)

early_stopper = EarlyStopper()
for i in range(num_epochs):
    print(f"Epoch {i+1}\n---------------------------------------------")
    train_loop(train_dataloader, model, loss_fn, optimizer)
    test_loop(test_dataloader, model, loss_fn)
    if val_loop(val_dataloader, model, loss_fn, early_stopper):
        print(f"Early stopping.")
        break
    scheduler.step()

print("Model trained!")
```

Batch Normalization Layer

We add a batch normalization layer in each of our convolution blocks right before the ReLU activation. As we saw in the "What Is Image Classification" section, the normalization is carried out differently during training versus inference – the mean and standard deviations are calculated for each mini-batch in training, whereas during inference, the moving averages of the mean and standard deviation accumulated during training are used for normalization. For this reason, to inform the model whether we are in the training or inference mode, we call model.train() in our train_loop, whereas we specify model.eval() in our test and validation loops.

Dropout Layer

We add a dropout layer after the ReLU activation of each of our convolution layers, with the hyperparameter p set to 0.2. As discussed before, each such dropout layer will drop each activation randomly with a probability of 0.2.

Here is a fourth version of our CNN, with the addition of batchnorm and dropout layers:

```
class ConvNeuralNetwork4(nn.Module):
    def __init__(self):
        super().__init__()
        # First convolution-block
        self.conv1 = nn.Conv2d(in_channels=1, out_channels=32, kernel_size=(3,3), padding=1)
        self.batch_norm1 = nn.BatchNorm2d(32)
        self.relu_act1 = nn.ReLU()
        self.dropout1 = nn.Dropout(p=0.2)
        self.max_pool1 = nn.MaxPool2d(kernel_size=(2,2))

        # Second convolution-block
        self.conv2 = nn.Conv2d(in_channels=32, out_channels=64, kernel_size=(3,3), padding=1)
        self.batch_norm2 = nn.BatchNorm2d(64)
        self.relu_act2 = nn.ReLU()
        self.dropout2 = nn.Dropout(p=0.2)
        self.max_pool2 = nn.MaxPool2d(kernel_size=(2,2))

        # Third convolution-block
        self.conv3 = nn.Conv2d(in_channels=64, out_channels=128, kernel_size=(5,5), padding=2)
        self.batch_norm3 = nn.BatchNorm2d(128)
        self.relu_act3 = nn.ReLU()
        self.dropout3 = nn.Dropout(p=0.2)
        self.max_pool3 = nn.MaxPool2d(kernel_size=(2,2))

        self.flatten = nn.Flatten()
        self.fully_connected = nn.Linear(3*3*128, 10)
```

```python
def forward(self, x):
    x = self.conv1(x)
    x = self.batch_norm1(x)
    x = self.relu_act1(x)
    x = self.dropout1(x)
    x = self.max_pool1(x)

    x = self.conv2(x)
    x = self.batch_norm2(x)
    x = self.relu_act2(x)
    x = self.dropout2(x)
    x = self.max_pool2(x)

    x = self.conv3(x)
    x = self.batch_norm3(x)
    x = self.relu_act3(x)
    x = self.dropout3(x)
    x = self.max_pool3(x)

    x = self.flatten(x)
    x = self.fully_connected(x)
    return x
```

This gives an accuracy of 99.4% on the test data, early stopping after just 22 epochs. Also, it is worth noting that it already gave a 98.2% accuracy after just one epoch for my run.

Optimizer Improvements

Apart from these, several improvements are possible to the optimizer, like using the momentum parameter with the optimizer, using a more sophisticated optimizer like Adam, etc. Pytorch facilitates all these options quite easily using its torch.optim library. However, we skip these details here and refer the reader to this documentation to further polish their model optimization process.

A More Challenging Dataset

Most of our code was generic and can be reused for any other dataset, with the exception of our model architecture code. The model architecture code used the dimensions of the input images to compute certain layer dimensions and will have to be adapted for a new dataset. We now pick the more difficult Oxford-IIIT Pet dataset and use a similar CNN to classify it.

Data Loading and Augmentation

To compensate for the low volume of data available, we use some data augmentation techniques implemented by transforms. PyTorch facilitates data augmentation through its torchvision.transforms module, which provides several different transforms. Different transforms can be chained together using the Compose function as we do below. These transformations are applied independently in each epoch on every batch so that in each epoch, the model sees a different transformed variation of each image. The original datasets are not modified by these transforms. We first define the transformations to be applied to the training data.

```
from torchvision.transforms import v2
import torchvision.transforms as transforms

train_transform = transforms.Compose([
      transforms.RandomResizedCrop(size=(224,224), scale=(0.6, 1.0)),
      transforms.RandomHorizontalFlip(p=0.4),
      transforms.RandomVerticalFlip(p=0.1),
      transforms.RandomRotation(10),
      transforms.ColorJitter(brightness=0.2, contrast=0.1,
      saturation=0.2, hue=0.1),
      transforms.ToTensor(),
      transforms.Normalize(mean=[0.485,
 0.456, 0.406], std=[0.229, 0.224, 0.225])
   ])
```

We have applied the following transforms in the given order:

1. **RandomResizedCrop**: This randomly crops the image within the specified scale and then resizes it to the specified size of (224,224).

2. **Flips**: We randomly flip the image horizontally with a probability of 0.4, and then randomly flip it vertically with a probability of 0.1.

3. **RandomRotation**: The image is then rotated randomly by up to ten degrees.

4. **ColorJitter**: We apply the color jitter, which might perturb its brightness, contrast, saturation, and hue by the specified factors (refer to this documentation to find out how the brightness_factor, etc., are calculated).

5. **ToTensor**: The image is converted to a tensor as before using the ToTensor transform.

6. **Normalization**: The images are then also normalized using ImageNet's mean and standard deviation, which is a common practice since ImageNet comprises millions of images that are representative of most real-world images. Alternatively, you can also calculate the mean and standard deviations for your own dataset and use those to normalize.

We then define the transforms to be applied to the test data, which just involves resizing, conversion to tensors, and normalization:

```
test_transform = transforms.Compose([
        transforms.Resize(size=(224,224)),
        transforms.ToTensor(),
        transforms.Normalize(mean=[0.485,
 0.456, 0.406], std=[0.229, 0.224, 0.225])
    ])
```

Data is loaded just like before, with the transforms we just defined:

```
train_data = datasets.OxfordIIITPet(
    root="data",
    split='trainval',
```

```
        download=True,
        target_types='category',
        transform=train_transform
)

test_data = datasets.OxfordIIITPet(
        root="data",
        split='test',
        download=True,
        target_types='category',
        transform=test_transform
)
```

We now print the data statistics:

```
print(train_data)
Dataset OxfordIIITPet
    Number of datapoints: 3680
    Root location: data
    StandardTransform
Transform: Compose(
               RandomResizedCrop(size=(224, 224), scale=(0.6,
               1.0), ratio=(0.75, 1.3333), interpolation=bilinear,
               antialias=True)
               RandomHorizontalFlip(p=0.4)
               RandomVerticalFlip(p=0.1)
               RandomRotation(degrees=[-10.0, 10.0], interpolation=nearest,
               expand=False, fill=0)
               ColorJitter(brightness=(0.8, 1.2), contrast=(0.9, 1.1),
               saturation=(0.8, 1.2), hue=(-0.1, 0.1))
               ToTensor()
               Normalize(mean=[0.485, 0.456, 0.406], std=[0.229,
               0.224, 0.225])

print(test_data)
Dataset OxfordIIITPet
    Number of datapoints: 3669
```

CHAPTER 3   IMAGE CLASSIFICATION USING CONVOLUTIONAL NEURAL NETWORKS

```
    Root location: data
    StandardTransform
Transform: Compose(
            Resize(size=(224, 224), interpolation=bilinear, max_
            size=None, antialias=True)
            ToTensor()
            Normalize(mean=[0.485, 0.456, 0.406], std=[0.229,
            0.224, 0.225])
          )
```

Note that the training data only has 3680 samples, with 37 categories. This makes image classification particularly challenging for this dataset. The data is fairly balanced and has about 100 samples per class, as indicated by the histogram in Figure 3-8.

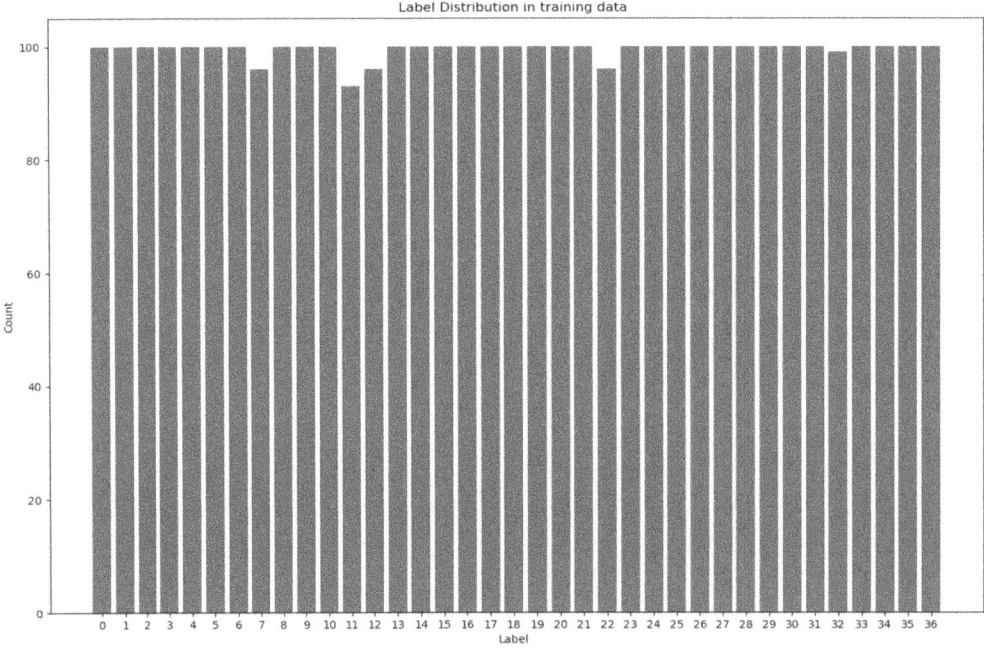

Figure 3-8. *Label counts for the Oxford-IIIT Pet dataset*

Below (Figure 3-9) are some randomly selected images from the dataset, along with their labels, which are integers corresponding to certain pet species. You can make out some cats and dogs in the images, although the lighting makes it pretty difficult to recognize some of them even for the human eye.

CHAPTER 3 IMAGE CLASSIFICATION USING CONVOLUTIONAL NEURAL NETWORKS

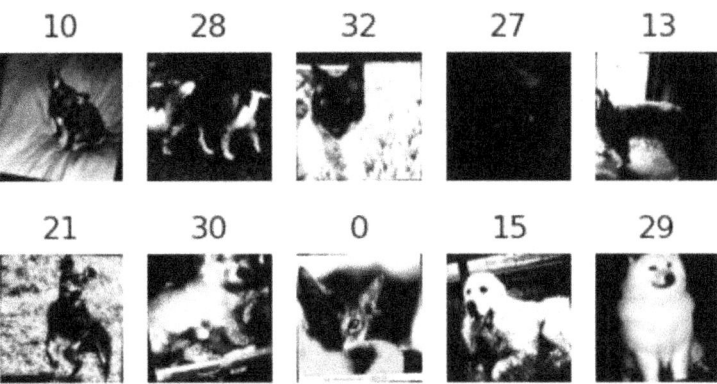

Figure 3-9. *Some sample images from the Oxford-IIIT Pet dataset*

Here is our fourth version of the CNN model that we defined before, with a few modifications for layer dimensions:

```
class ConvNeuralNetwork4(nn.Module):
    def __init__(self):
        super().__init__()
        # First convolution-block
        self.conv1 = nn.Conv2d(in_channels=3, out_channels=32, kernel_
        size=(3,3), padding=1)
        self.batch_norm1 = nn.BatchNorm2d(32)
        self.relu_act1 = nn.ReLU()
        self.dropout1 = nn.Dropout(p=0.2)
        self.max_pool1 = nn.MaxPool2d(kernel_size=(2,2))

        # Second convolution-block
        self.conv2 = nn.Conv2d(in_channels=32, out_channels=64, kernel_
        size=(3,3), padding=1)
        self.batch_norm2 = nn.BatchNorm2d(64)
        self.relu_act2 = nn.ReLU()
        self.dropout2 = nn.Dropout(p=0.2)
        self.max_pool2 = nn.MaxPool2d(kernel_size=(2,2))

        # Third convolution-block
        self.conv3 = nn.Conv2d(in_channels=64, out_channels=128, kernel_
        size=(5,5), padding=2)
```

```
        self.batch_norm3 = nn.BatchNorm2d(128)
        self.relu_act3 = nn.ReLU()
        self.dropout3 = nn.Dropout(p=0.2)
        self.max_pool3 = nn.MaxPool2d(kernel_size=(2,2))

        self.flatten = nn.Flatten()
        self.fully_connected = nn.Linear(28*28*128, 37)
    def forward(self, x):
        x = self.conv1(x)
        x = self.batch_norm1(x)
        x = self.relu_act1(x)
        x = self.dropout1(x)
        x = self.max_pool1(x)

        x = self.conv2(x)
        x = self.batch_norm2(x)
        x = self.relu_act2(x)
        x = self.dropout2(x)
        x = self.max_pool2(x)

        x = self.conv3(x)
        x = self.batch_norm3(x)
        x = self.relu_act3(x)
        x = self.dropout3(x)
        x = self.max_pool3(x)

        x = self.flatten(x)
        x = self.fully_connected(x)
        return x
```

This fairly simple model gives an accuracy of a measly 7.7% on my machine. In fact, a uniform random prediction would give an accuracy of about 1/37*100 = 2.702%. So our rudimentary model is only slightly better than a random prediction. We would have to put more work into our architecture to get a better performance. Instead of building a deep architecture from scratch, we try our hand at transfer learning using a pretrained model in the next section.

Transfer Learning: Fine-Tuning a Pretrained Model

Since the Oxford-IIIT pet is a challenging dataset, we might need to significantly improve the architecture to give a reasonable accuracy. Moreover, as we saw, the small volume of training data is what makes this dataset particularly challenging – so even a much better architecture might have limitations on how well it could work with such a small quantity of training data. To address these concerns, we use a technique called transfer learning. On a high level, transfer learning corresponds to transferring the skills that the model picked up on some other (potentially better and larger) dataset to the current dataset, with a few adjustments as needed. This eliminates the need for the model to learn from scratch, which, as we saw in the last section, could be challenging with a low data volume.

In the context of deep learning, this involves starting with models pretrained on large volumes of data and relearning the parameters for a few (or all) layers by training for a few epochs.

PyTorch offers a plethora of powerful pretrained models for ready use. We shall work with the resnet18 model, which is a residual neural network model, which means it has some skip connections inside the CNN. It has 18 layers, which is where it gets its name. We shall skip the rest of the details of a residual neural network here since they would be out of the scope of this book.

We now load the resnet18 model from the torchvision.models module:

```
from torchvision import models
resnet = models.resnet18(pretrained=True)
print(resnet)
```

This prints the entire architecture of the loaded model, which we shall not display here to avoid clutter. Do run it for yourself and look through this architecture to get an idea of it. You will be able to make sense of it all from the concepts we have seen so far. It uses the same building blocks that we used, like Conv2d, BatchNorm2d, ReLU, MaxPool2d, etc. The only main difference is that this network is much deeper and hence computationally very expensive to train. This has been trained using the huge ImageNet dataset.

We just print the last few layers here to get a sense of the sizes of layers:

```
(layer4): Sequential(
    (0): BasicBlock(
```

```
    (conv1): Conv2d(256, 512, kernel_size=(3, 3), stride=(2, 2),
    padding=(1, 1), bias=False)
    (bn1): BatchNorm2d(512, eps=1e-05, momentum=0.1, affine=True, track_
    running_stats=True)
    (relu): ReLU(inplace=True)
    (conv2): Conv2d(512, 512, kernel_size=(3, 3), stride=(1, 1),
    padding=(1, 1), bias=False)
    (bn2): BatchNorm2d(512, eps=1e-05, momentum=0.1, affine=True, track_
    running_stats=True)
    (downsample): Sequential(
      (0): Conv2d(256, 512, kernel_size=(1, 1), stride=(2, 2),
      bias=False)
      (1): BatchNorm2d(512, eps=1e-05, momentum=0.1, affine=True, track_
      running_stats=True)
    )
  )
  (1): BasicBlock(
    (conv1): Conv2d(512, 512, kernel_size=(3, 3), stride=(1, 1),
    padding=(1, 1), bias=False)
    (bn1): BatchNorm2d(512, eps=1e-05, momentum=0.1, affine=True, track_
    running_stats=True)
    (relu): ReLU(inplace=True)
    (conv2): Conv2d(512, 512, kernel_size=(3, 3), stride=(1, 1),
    padding=(1, 1), bias=False)
    (bn2): BatchNorm2d(512, eps=1e-05, momentum=0.1, affine=True, track_
    running_stats=True)
  )
)
(avgpool): AdaptiveAvgPool2d(output_size=(1, 1))
(fc): Linear(in_features=512, out_features=1000, bias=True)
)
```

As we can see, the last layer is a fully connected linear layer with 512 inputs and 1000 outputs. The ImageNet dataset has 1000 classes, which explains the 1000 outputs. We will replace this last layer with a layer with 37 outputs as suits our dataset.

We now set all the hyperparameters, along with the optimizer, etc.:

```python
from torch.optim.lr_scheduler import ExponentialLR
torch.manual_seed(9) # To reduce variability in each run

# Hyperparameters
learning_rate = 0.1
batch_size = 64
num_epochs = 40

#Defining model, loss, optimizer
loss_fn = nn.CrossEntropyLoss()

model = models.resnet18(pretrained=True) # Loading resnet18
for param in model.parameters(): # freezing all the layers
    param.requires_grad = False

# Replacing the last dense layer to have 37 outputs
model.fc = nn.Linear(512, 37, bias=True)

optimizer = torch.optim.SGD(
    filter(lambda p: p.requires_grad, model.parameters()),
    lr=learning_rate, momentum=0.8
)

scheduler = ExponentialLR(optimizer, gamma=0.9)
early_stopper = EarlyStopper()
```

Finally, we run the training, validation, and testing loops as before:

```python
for i in range(num_epochs):
    print(f"Epoch {i+1}\n-------------------------------------------")
    train_loop(train_dataloader, model, loss_fn, optimizer)
    test_loop(test_dataloader, model, loss_fn)
    if val_loop(val_dataloader, model, loss_fn, early_stopper):
        print(f"Early stopping.")
        break
    scheduler.step()

print("Model trained!")
```

CHAPTER 3 IMAGE CLASSIFICATION USING CONVOLUTIONAL NEURAL NETWORKS

Since requires_grad is set to False for all but the last layer, these other layers will not update their parameters during the training loop. These layers are said to be frozen. The training loop is mainly aimed at learning the parameters for the last dense layer, which we modified.

This fine-tuned model gave an accuracy of about 83.2%, significantly better than that of our handcrafted model or a random prediction. You can try more complex models like ResNet-50 (which has 50 layers instead of 18 as in resnet18) or try to fine-tune more layers of some of these models instead of just the final layer as we did, to get better accuracy.

Summary

- Image classification is the problem of classifying input images into one of several predefined classes.

- Convolutional neural networks have been outperforming all other techniques on image classification tasks since 2012.

- Convolutions are a mathematical operation applied to images, or more generally, to feature maps. They involve filters or kernels containing weights that slide over the image and take the sum of element-wise products of these weights with the pixel values, which go into the corresponding position in the 2D output.

- A convolution layer comprises a bunch of convolution operations applied to the input feature maps, followed by activations such as ReLU. It is common practice to pad the inputs, usually with 0s, before applying the convolution to maintain the input width and height.

- A pooling layer reduces the dimension of its input by applying an aggregate function on adjacent input activations. For example, a 2×2 max pool layer considers a partition of its input into 2×2 submatrices and takes the max over each of them.

- After applying a series of convolution and pooling layers, a Flatten layer is applied, which flattens the 3D feature map into a 1D vector. This is typically followed by one or more dense layers, followed by softmax.

- Softmax is a way of converting a vector of numerical scores into probabilities.

- We typically use cross-entropy loss or log loss for CNNs, which is a way of measuring the dissimilarity between two distributions.

- Batch normalization layers are applied after the convolution operation and before the activation. It helps in regularization and improves the training process in various ways.

- Dropout layers are applied after the activation functions as a way of regularization. They randomly zero out each activation with some probability p.

- Data augmentation is the process of perturbing the training data so as to expose the model to more variations of the data to make it more generalizable. For image classification, some common data augmentation techniques include flipping or rotating the images; changing the brightness, contrast, etc.; or randomly cropping the images.

- The main components of an image classification project typically involve loading the datasets into data loaders – which help in iterating over the dataset in batches (with or without shuffling), data exploration and visualization, defining the model architecture, and writing and executing the model training, validation, and testing loops.

- Any neural network model you write in PyTorch should usually be subclassed from the torch.nn module. It should include the __init__ function containing the layer definitions and the forward() function, which is executed during the forward pass.

- Calling backward() on the loss function populates the .grad attributes of all the model parameters, i.e., the layer weights.

- The training process can be improved by using a learning rate scheduler, which gradually reduces the learning rate as the model trains, and by using early stopping, which helps avoid overfitting by stopping the learning process if the validation metrics stop showing significant improvement.

- For datasets with a low volume of training data, transfer learning is a useful technique. Transfer learning involves loading a CNN model pretrained on a larger corpus of data and retraining some or all of its layers using the current dataset. This process is also called fine-tuning the pretrained model, and the layers whose parameters we don't retrain are said to be frozen.

CHAPTER 4

Introduction to Natural Language Processing: Building a Text Classifier

From this chapter, we start the second major area of AI, namely, natural language processing (NLP). We begin the chapter by giving a primer on NLP. We mention the different types of problems addressed by NLP. The chapter then goes on to describe the different steps of a typical NLP project, including converting text data into numerical data, one-hot encodings, and word embeddings. We then describe the encoder-decoder architecture and some of the popular and frequently used neural network approaches to implementing them, like RNNs, LSTMs, and transformers.

The third section dives into the project of text classification, which is the main focus of this chapter. It starts by describing the dataset used and then goes into the details of data processing pipelines, the model architecture, training and testing loops, etc. We then describe our second model, which is based on pretrained embeddings generated by the Word2Vec model. Finally, we use a pretrained version of the popular transformer model DistilBERT to produce dense vector representations, which are then used to train our classifier. We load the DistilBERT model from the Hugging Face Hub, so this section also provides a short tutorial on loading and using Hugging Face models.

CHAPTER 4 INTRODUCTION TO NATURAL LANGUAGE PROCESSING: BUILDING A TEXT CLASSIFIER

Natural Language Processing (NLP)

What Is NLP

Natural language processing (NLP), as the name suggests, involves computers learning to process, understand, analyze, and generate "natural" or human language (as opposed to formal languages or programming languages). It has been a burgeoning area of research since almost 1950s. It has become ubiquitous in popular media recently because of text generation services like ChatGPT, which could achieve incredible feats, ranging from passing various medical and law exams and solving math Olympiad questions to writing fluent poetry and essays on any topic. This has been possible because of the more recent revolution in the field brought by the transformer architecture.

Popular Problems Solved by NLP

Here are some prominent problems that NLP focuses on:

1. **Machine Translation**: A major focus area where text is translated from one language to another, since as early as the 1950s

2. **Text Summarization**: Summarizing a document, book, research article, etc.

3. **Code Generation**: Writing code based on instructions in English or as pseudocode

4. **Text Generation**: Writing poetry, essays, etc.

5. **Dialogue Systems**: Used in chatbots

6. **Text Classification**: Sentiment analysis of reviews, categorizing news articles, etc.

7. **POS Tagging**: Parts of speech tagging in sentences

8. **Question Answering**: Answering questions based on a provided text excerpt

Preprocessing

You may have noticed that all the ML models we have seen so far (in the area of computer vision) accepted fixed-length numerical inputs. This might make you wonder: How do we convert the variable-length raw text input into a fixed-length numerical input while maintaining the essence of the text? The following preprocessing steps will guide us in that direction.

Input Encoding: Converting Text into Numbers

We first convert our text into a series of numbers via a process called *encoding*. Encoding involves two steps: tokenization and numericalization.

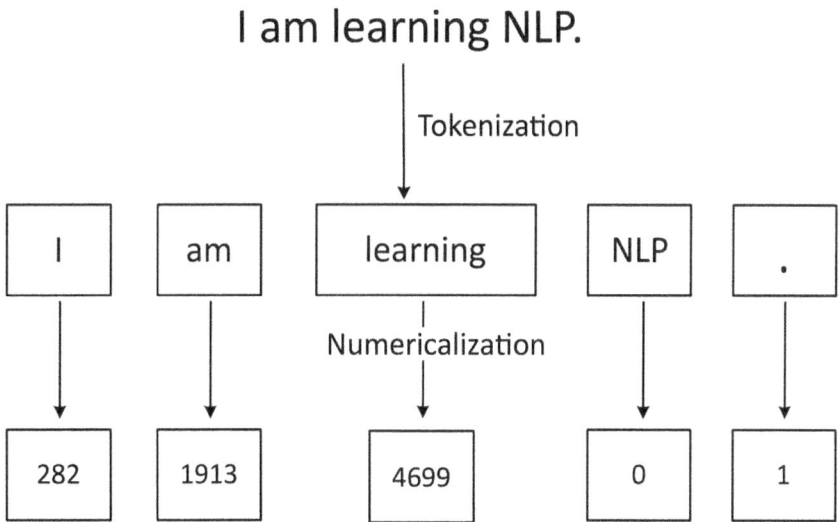

Figure 4-1. *Encoding: tokenization + numericalization*

Tokenization

The first step is to split a text into an array of tokens. So a text input string like "I am learning NLP." is converted to the list ["I", "am", "learning", "NLP", "."]. Now, at least we have a list of discrete tokens (i.e., words) to work with. This specific technique of splitting the text into a list of words is called *word tokenization*. Other options are sentence tokenization or character tokenization, where each sentence or each character is considered a token, respectively. The most popular technique currently is subword tokenization, which involves breaking each word into commonly occurring subwords. We will not go into the details of subword tokenization in this book.

Numericalization

These discrete tokens are still symbols, whereas our ML models understand numbers. The process of numericalization converts this list of tokens into a list of numbers. Each token is mapped to a unique (usually non-negative) integer ID or index. The entire token list is thereby mapped to a list of indices. We typically build a vocabulary mapping by first going through the entire text and assigning a unique integer index to each word in the text (and also each punctuation). A special index is reserved for unknown tokens, so that any unknown token not found in the training dataset is mapped to this special index.

In this way, tokenization followed by numericalization converts text into an array of non-negative integers. This entire process is called input encoding, as depicted in Figure 4-1.

Truncation and Padding

Input encoding yields sequences that are numerical, but still have variable lengths. Most ML models are optimized to process inputs batchwise, where each batch is expected to have inputs of the same length. For example, if you are working with PyTorch, each batch is typically a PyTorch tensor where each row contains an input sequence. PyTorch requires every row to be of the same length, and as we saw in Chapter 2, it can execute fast tensor operations because this assumption allows it to have its data stored in contiguous blocks of physical memory. To ensure uniform sequence lengths within each batch, shorter sequences are padded by a fixed value, usually to the right.

Some ML models also put an upper limit on the number of tokens each sequence can contain. This constraint can be met by simply truncating sequences that are too long.

One-Hot Encodings and Embeddings

One-Hot Encodings

Having an index for each word is a good start toward converting text into numerical inputs. However, since these indices are chosen arbitrarily (perhaps in the order in which these words were encountered in the text), this sets up an artificial ordering among the word inputs. For example, the word corresponding to index 10 might be "gorilla," the one corresponding to index 11 might be "sky," and the one having index 3001 might be "chimpanzee." This sends a false signal to the model that a "gorilla" is closer to "sky" semantically than it is to a "chimpanzee."

To avoid this false ordering among words, a common technique is to convert our index-based encodings into one-hot encoded vectors. This involves using a binary vector of the size of our vocabulary for each word. The vector representing any given word has a "1" at the index of the encoding of that word and a 0 everywhere else. For example, if "banana" had index 10, then its one-hot encoded vector would have a 1 at the tenth position and 0 everywhere else.

Embeddings

Using one-hot vectors causes the input dimension to blow up to the size of the vocabulary. Moreover, it does not capture any similarity in semantics at all. An embedding layer is therefore used to convert the one-hot encoded vectors into embedded vectors. These vectors capture some semantic similarities of words. For example, the distance between the embedded vectors of "gorilla" and "chimpanzee" would be smaller compared to that between the vectors for "gorilla" and "sky."

Text Classification

We shall now study some state-of-the-art NLP model architectures. Although we shall look at these in the context of text classification, these architectures are generic and are applicable to almost all the NLP tasks we mentioned before.

A Typical NLP Modeling Approach

The primary approach in almost all modern NLP deep learning frameworks is to create a rich numerical representation of the input text and then use this representation as input to any ML model (typically neural networks) for classification (or any other task).

We start with input encodings described earlier, which are fixed mappings of the text to some integer indices or, equivalently, one-hot encoded vectors. This is a bland way of representing words since it doesn't capture any information about the meanings or the relative context of words. All the models that we shall see aim to convert these basic representations of words into more meaningful dense vector representations so as to capture the meaning as well as context of each word in a sentence. Subsequently, these word vectors can be pooled in some way – by taking the mean, max, or sum – to get a vector capturing the essential meaning of a sentence or even a paragraph or a document.

The Encoder Architecture

The term encoder is overloaded with multiple, although related, meanings in NLP. We saw the preprocessing step of input encodings before. In this context, however, an encoder is a model architecture that converts a variable-length input sequence (typically input encodings of a bunch of sentences) into a vector representation of some fixed dimension (see Figure 4-2). This vector can then be passed through any classifier to predict from a specified set of target classes.

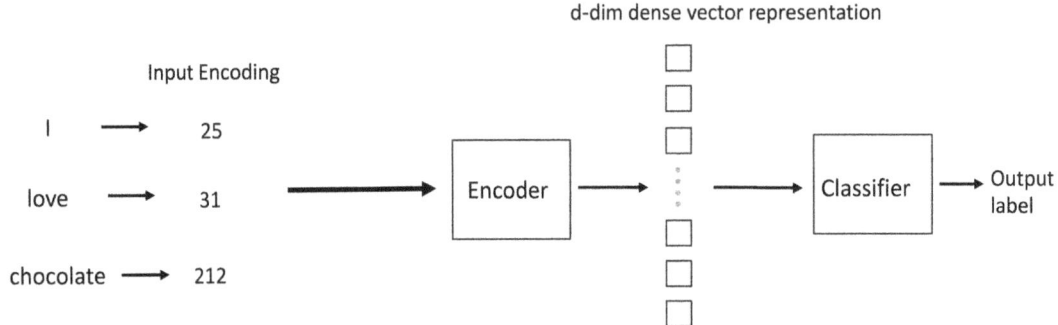

Figure 4-2. Encoder architecture for classification

So we need to train both the encoder model and the classifier. In many cases, both of these are neural networks and can therefore be trained together using gradient descent. This means that the forward pass will begin at an input sequence, going through the encoder, followed by the classifier to predict some prediction (or prediction probability distribution). The cross-entropy loss (log-loss) will be used as the loss function, and the backward pass will propagate the gradients through the classifier to update the classifier parameters and then through the encoder to update the encoder parameters as well.

The classifier is sometimes also termed the "classification head," especially when it is a single-layer feed-forward neural network.

Types of Model Training

Large companies and research groups train powerful encoders with billions of parameters on large amounts of data. This process is compute intensive and may not be feasible for individual users. Some groups have been generous enough to put some of their pretrained models online, to be freely used by any interested users.

Based on whether a pretrained encoder is used as is, or trained along with the classifier, three cases arise:

1. **Pretrained Encoders**: The training data used to train these encoders is often quite diverse, so it might be worthwhile to use these pretrained encoders as is while training just the classification head. In this case, the pretrained model acts as a *feature extractor*, while the classifier uses these features for the task at hand.

2. **Fine-Tuning Encoders**: Another alternative is to start with a pretrained encoder and use it merely as an initialization for the parameters. We then continue to train both the encoder and the classifier using our training data in the same backward pass. We can train some or all of the parameters of the pretrained encoder model in this way. This process is called fine-tuning the encoder.

3. **Building from Scratch**: Alternatively, we can start from scratch with an encoder as well as a classifier, with randomly initialized parameters, and train them both together. This would typically be the case if you have a lot of compute power to train powerful models and if you have plentiful data for training.

RNNs and LSTMs (Optional)

To implement an encoder, your first thought might be to build neural networks that accept a fixed input length, say 10, 20, or 100 words at a time. However, this fixed window model immediately puts a limitation on the input size, along with other limitations that we won't go into.

This fixed-size limitation is circumvented by using what are called recurrent neural networks, also popularly known as RNNs, to implement encoders. On a high level, an RNN is a single-layer neural network with a "feedback loop." In other words, at each time step t, an RNN produces an output o_t and also another vector called the hidden state h_t. This hidden state h_t is fed back to the model at time step t+1, along with the (t+1)th input token. In this way, the RNN can be fed inputs one token at a time, and at any time step, it is able to retain some memory of the previous tokens.

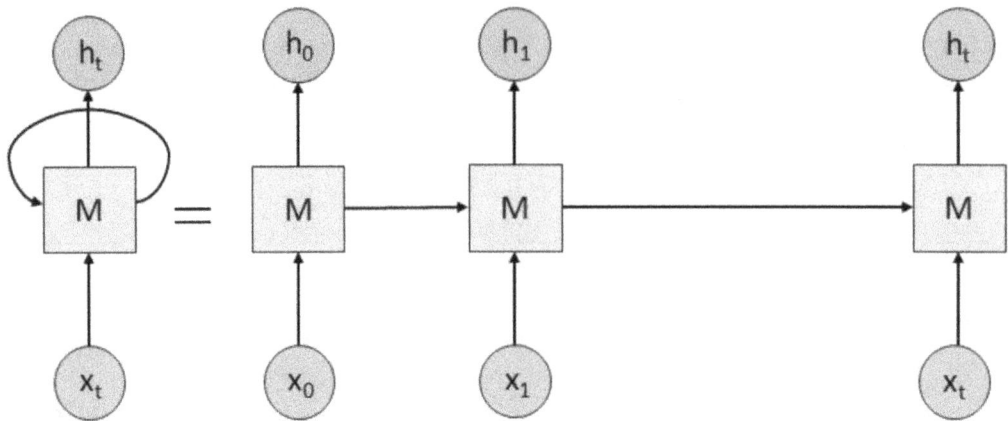

Figure 4-3. RNN architecture

As depicted in Figure 4-3, you can understand an RNN better by unfolding it with time steps. It then looks like a long chain of feed-forward neural networks, where each network feeds into the next one, and they all have the same parameters. This is the order in which computation proceeds in the forward pass. The backward pass backpropagates the gradient in the exact reverse order.

RNNs can be used as encoders to obtain dense vector representations of input text and also as decoders to generate new text.

One drawback of RNNs is that they can only remember so much of the historical context. It is observed in practice that only a few of the most recent tokens are usually remembered. Moreover, they also often suffer from training problems, like the vanishing gradient and the exploding gradient problems. Both these issues are overcome by the Long Short-Term Memory (LSTM) architecture. In essence, LSTMs contain additional gates that control which information can be read, written, and erased from their memory. LSTMs are essentially RNNs with a longer short-term memory, hence the name.

These architectures have a drawback in that they only look at the left context, i.e., words to the left of the word being processed. To benefit from getting both contexts, bidirectional RNNs (or LSTMs) are used. These are essentially a combination of two RNNs (LSTMs), one that processes from left to right and another that processes tokens from right to left. Finally, the hidden states obtained from both these RNNs (LSTMs) are concatenated together.

The Transformer Architecture

The transformer is by far the most popular choice for implementing encoders as well as decoders. Cutting-edge GenAI systems like ChatGPT are based off of transformers.

Motivation

RNNs/LSTMs have some downsides:

1. While processing a word, the influence of other words near it is much greater than those that are farther away. For example, in a sentence like "The person, who is wearing blue shoes and a red hat, is running.", the model might not easily be able to infer that it is "the person" who "is running".

2. Words have to be processed sequentially, which makes the model slower.

Transformers, proposed in the paper "Attention Is All You Need" by Vaswani et al., try to overcome these limitations by introducing the self-attention mechanism.

Self-attention Mechanism

Self-attention is the most important building block of the transformer architecture. The main idea is that in each self-attention block, a word's representation keeps getting enriched by its contextual surrounding words that are most relevant to it.

We see a high-level view of how this is achieved:

1. Each word representation is projected into three vectors: a query, a key, and a value.

2. The attention score of a word w is computed with every other word as follows: the query of w is compared to every key by taking their dot product and then scaled down by sqrt(d), where d is the dimension of the key.

3. Softmax is used to normalize these attention scores to give attention weights. So now, for this word that we are processing, we have a similarity score or weight between 0 and 1 with every other word in our sequence (including itself).

4. These attention weights are then used to weigh the value vectors corresponding to each word, and the representation for our word in focus is updated by this weighted sum of value vectors.

In this way, the representation of a word becomes a carefully calculated concoction of all the other words in the sequence.

We can understand this better with an example. Consider the sentence:

"Little Emma stood by the kitchen counter munching on a yellow mango which she spilled on her new red dress."

A good model should comprehend, among other things, that

1. Yellow refers to the color of the mango.

2. Red applies to the dress.

3. The pronoun "her" refers to Emma, as does the adjective "Little".

4. And that Emma is the one munching on the mango (and not, for instance, the "kitchen counter" even though those words are nearer to munching in the sentence).

Self-attention does this by processing each word (in parallel), starting with computing all the keys, queries, and values[1]. The model will process the word "munching" with the following steps:

1. The model will look up from its scratchpad the query corresponding to "munching".

2. It will look up all the keys from its scratchpad and compare the query with each key using dot products.

---

[1] We shall skip these details here, but it is essentially just taking a linear projection by multiplying the word vector by a matrix.

3. It might find, for example, that munching is 10% similar to mango (i.e., the query of munching is 10% similar to the key of mango), 2% similar to kitchen, 15% similar to Emma, and 73% similar to munching itself! (Yes, the query is also compared to its own key as well!)

4. It will compute a new representation for munching by weighing the value vectors by these similarities that it found, as

$$0.1\, v_{mango} + 0.02\, v_{kitchen} + 0.15\, v_{Emma} + 0.73\, v_{munching}$$

The representations of each word get refined in this way in each self-attention layer.

Transformer-Based Encoder Architecture

Transformer-based encoders are essentially comprised of self-attention blocks and feed-forward neural network blocks in alternation. There are a few additional details to chalk out here, as we discuss next.

Positional Embeddings

Our models described so far have no way of deciphering the relative position of each word. This is addressed by encoding the index of each word in the sequence as a positional embedding and adding it to the input.

Layer Normalization

Layer normalization is added before each layer. It normalizes the inputs to each layer and helps the model with stability and generalization.

The Bidirectional Encoder Representations from Transformers (BERT) model is a prime example of a transformer-based encoder.

Sequence-to-Sequence Models

Decoder

We saw the encoder architecture, which aims at encoding a text sequence into a fixed-sized dense vector representation. For tasks like classification, which require a numerical output, these encoded representations can readily be passed through classification models to predict an output label.

However, some tasks, such as machine translation, summarization, etc., require a sequence of text to be produced as output. The decoder architecture does this job of converting an encoded state, i.e., dense vector representation, of the input to an output text sequence. Decoders can be implemented via RNNs (or LSTMs) as well as transformers.

Encoder-Decoder

This combination of encoder and decoder can be used together to first convert the input text to a fixed-dimensional hidden state and then convert it back from the hidden state to the desired format of output text. Note that this output text could be in a different language compared to the input text, as is the case for machine translation.

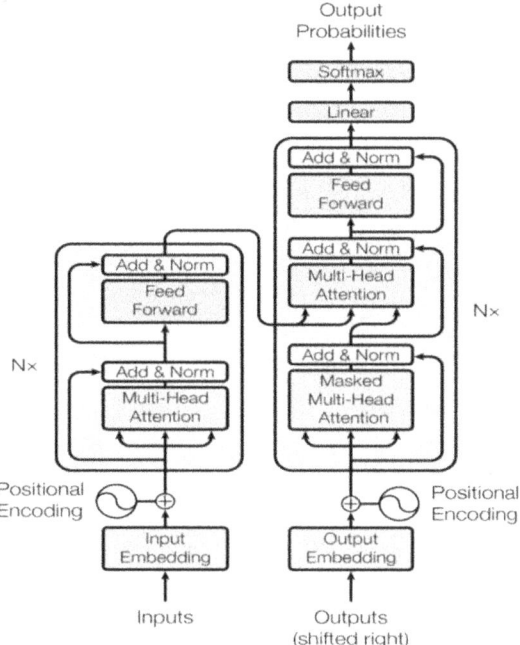

Figure 4-4. *Picture from the paper "Attention Is All You Need"*

Ref: https://proceedings.neurips.cc/paper_files/paper/2017/file/3f5ee243547dee91fbd053c1c4a845aa-Paper.pdf

This encoder-decoder architecture (Figure 4-4) is therefore a way to convert an input sequence into an output sequence, and any model that implements this is termed a sequence-to-sequence model.

The Text Classification Project

Let us now get started with our first NLP project. As we saw before, text classification involves classifying an input text into one of a few predefined classes.

We will be working with a dataset containing short news articles labelled with a category from one of four fixed categories: world, sports, business, and science and technology.

CHAPTER 4 INTRODUCTION TO NATURAL LANGUAGE PROCESSING: BUILDING A TEXT CLASSIFIER

Imports

We will need the following general imports to be able to run the rest of our code:

```
import torch
from torch.utils.data import Dataset
import matplotlib.pyplot as plt
from torch import nn
from torchtext import datasets
from torch.utils.data import random_split
import matplotlib.pyplot as plt
```

We will require more imports relevant to specific parts of the code, which we shall mention as needed.

Dataset

We shall work with the AGNews dataset, which is a part of torchtext.datasets. This dataset can be accessed as

```
from torchtext.datasets import AG_NEWS
from torchtext.data import to_map_style_dataset

train_set = AG_NEWS(split="train")
train_iter = iter(train_set)
train_set_map = to_map_style_dataset(train_set)

test_set = AG_NEWS(split="test")
test_iter = iter(test_set)
test_set_map = to_map_style_dataset(test_set)
```

We convert train_set and test_set to both an iterable-style dataset and a map-style dataset. In short, an iterable-style dataset in PyTorch facilitates direct iteration over the data, but doesn't allow random access via indexing. However, it makes it more efficient to iterate over large datasets. In contrast, a map-style dataset allows random (indexed) access to the data and makes preprocessing and batching easier.

CHAPTER 4   INTRODUCTION TO NATURAL LANGUAGE PROCESSING: BUILDING A TEXT CLASSIFIER

Dataset Statistics

When you start any project in ML, always take your time to explore the data first. Taking a closer look at various data statistics helps to get a sense of what you are dealing with and saves a lot of your time later when working on solutions.

We will start by finding out the sizes of the training and test sets:

```
# DATASET SIZES
print(f"Training set size: {len(train_set_map)}")
print(f"Test set size: {len(test_set_map)}")

# Unique labels
labels_set = {label for label, _ in train_set_map}
print(f"The set of unique labels: {labels_set}")

>>> Training set size: 120000
>>> Test set size: 7600
>>> The set of unique labels: {1, 2, 3, 4}
```

As you can see, it is a fairly large dataset with 120K training samples and 7600 test samples. There are four integer class labels 1-4, corresponding to the categories "World", "Sports", "Business", and "Science and Technology" in that order (this information is not evident from the dataset itself).

We print a few samples of each of the class types:

```
# Create a PyTorch tensor of labels
labels_list = [label for label, _ in train_set_map]
labels_tensor = torch.tensor(labels_list)

# Count the frequency of each label
unique_labels, counts = torch.unique(labels_tensor, return_counts=True)

label_dict = {1: "World", 2:"Sports", 3:"Business", 4: "Science & Technology"}

# PRINTING SOME SAMPLES FOR EACH LABEL
print_sample_size = 2
for label in unique_labels:
    count = 0
```

```
    print(f"Some samples with label {label}, corresponding to category
    \"{label_dict[int(label)]}\"")
    for l, t in train_set_map:
        if l == label:
            print(t)
            count += 1
            print(" ")
            if count >= print_sample_size:
                break
    print("----------------------------------------------------------------")
```

On my laptop, this gave me the output:

Some samples with label 1, corresponding to category "World"
Venezuelans Vote Early in Referendum on Chavez Rule (Reuters) Reuters - Venezuelans turned out early\and in large numbers on Sunday to vote in a historic referendum\that will either remove left-wing President Hugo Chavez from\office or give him a new mandate to govern for the next two\years.

S.Koreans Clash with Police on Iraq Troop Dispatch (Reuters) Reuters - South Korean police used water cannon in\central Seoul Sunday to disperse at least 7,000 protesters\urging the government to reverse a controversial decision to\send more troops to Iraq.

Some samples with label 2, corresponding to category "Sports"
Phelps, Thorpe Advance in 200 Freestyle (AP) AP - Michael Phelps took care of qualifying for the Olympic 200-meter freestyle semifinals Sunday, and then found out he had been added to the American team for the evening's 400 freestyle relay final. Phelps' rivals Ian Thorpe and Pieter van den Hoogenband and teammate Klete Keller were faster than the teenager in the 200 free preliminaries.

Reds Knock Padres Out of Wild-Card Lead (AP) AP - Wily Mo Pena homered twice and drove in four runs, helping the Cincinnati Reds beat the San Diego Padres 11-5 on Saturday night. San Diego was knocked out of a share of the NL wild-card lead with the loss and Chicago's victory over Los Angeles earlier in the day.

Some samples with label 3, corresponding to category "Business"
Wall St. Bears Claw Back Into the Black (Reuters) Reuters - Short-sellers, Wall Street's dwindling\band of ultra-cynics, are seeing green again.

Carlyle Looks Toward Commercial Aerospace (Reuters) Reuters - Private investment firm Carlyle Group,\which has a reputation for making well-timed and occasionally\controversial plays in the defense industry, has quietly placed\its bets on another part of the market.

Some samples with label 4, corresponding to category "Science & Technology"
'Madden,' 'ESPN' Football Score in Different Ways (Reuters) Reuters - Was absenteeism a little high\on Tuesday among the guys at the office? EA Sports would like\to think it was because "Madden NFL 2005" came out that day,\and some fans of the football simulation are rabid enough to\take a sick day to play it.

Group to Propose New High-Speed Wireless Format (Reuters) Reuters - A group of technology companies\including Texas Instruments Inc. (TXN.N), STMicroelectronics\(STM.PA) and Broadcom Corp. (BRCM.O), on Thursday said they\will propose a new wireless networking standard up to 10 times\the speed of the current generation.

We will also plot a histogram of the label distribution:

```
# Plotting the histogram for label distribution
plt.bar(unique_labels, counts,tick_label=list(range(1,5)))
plt.xlabel("Label")
plt.ylabel("Count")
plt.title("Label Distribution in training data")

plt.show()
```

CHAPTER 4 INTRODUCTION TO NATURAL LANGUAGE PROCESSING: BUILDING A TEXT CLASSIFIER

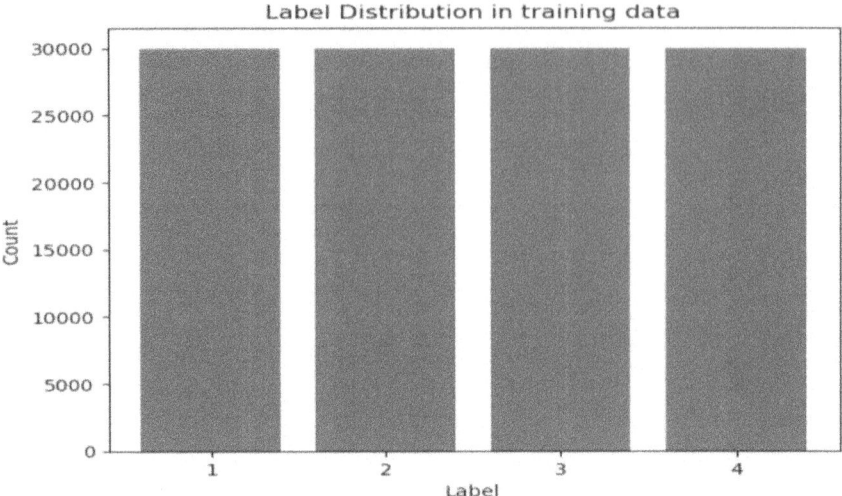

Figure 4-5. *Label counts in AGNews training data*

There is an equal distribution of samples among the four classes, as can be seen from the histogram (see Figure 4-5).

Preprocessing

Preprocessing is essentially the process of converting the raw text of each sample into a numerical format that can be readily ingested by an ML model.

Input Encoding

As we saw in the "Natural Language Processing (NLP)" section, we convert the raw text into a list of integer indices via a process called input encoding. Encoding involves tokenization followed by numericalization.

Tokenization

The first step is to convert the raw text, which is a string, into a list of suitable strings or *tokens*. We will work with word tokenization, where each word is a token. We could write a tokenizer by hand, but there are many excellent tokenizers already provided in different libraries.

A popular choice is the "basic_english" tokenizer, which we can load using the get_tokenizer function in the data.utils library of torchtext:

```
from torchtext.data.utils import get_tokenizer
tokenizer = get_tokenizer("basic_english")
```

Let us see the effect our tokenizer has on a sample sentence:

```
test_sentence = "I am learning NLP. I am loving it!!"
tokenizer(test_sentence)
>>> ['i', 'am', 'learning', 'nlp', '.', 'i', 'am', 'loving', 'it', '!', '!']
```

Three things can be noticed right away about this tokenizer:

1. It separates each word into a token.
2. It converts all the alphabet into lowercase (like I becomes "i" and NLP becomes "nlp").
3. Punctuations are treated as separate tokens (like "!" and ".").

Numericalization

Numericalization involves assigning a fixed integer for each token in our vocabulary. Therefore, the preparatory step toward numericalization is to first build a vocabulary from our training corpus.

We use the torchtext method build_vocab_from_iterator to build a vocab object from the sample text. For that, we first need to create a token_generator, which is essentially a generator in Python that yields a list of tokens corresponding to each text sample.

```
# Numericalization
from torchtext.vocab import build_vocab_from_iterator
token_generator = (tokenizer(text) for label,text in train_iter)
my_vocab = build_vocab_from_iterator(token_generator, specials=["<unk>"])
my_vocab.set_default_index(my_vocab["<unk>"])
```

A special token <unk> is added for words that are not in the vocabulary built from the training text, and a default index is set for this token. Whenever an unknown word is encountered during inference (or testing), our vocab object will map it to this default index. Conversely, when decoding, i.e., looking up tokens from a list of indices, the default index is mapped to the <unk> token, as we will see later.

The object my_vocab is of type Torchtext.vocab.Vocab. This contains a __get_item__ method, so it can answer queries by indexing, like

```
my_vocab["football"]
>>> 341
my_vocab["Hello"]
>>> 0
my_vocab.get_default_index()
>>> 0
```

As you might have guessed, 341 is the index of the word "football" in our vocabulary, whereas 0 is the index of the word "Hello". The default index is also 0, which means that our vocabulary hasn't been exposed to the word "Hello" yet, which might be because we are working with a news dataset.

A vocab object can be paired harmoniously with the tokenizer to yield a list of encodings:

```
encoded_input = my_vocab(tokenizer("I am learning NLP. I am loving it!!"))
print(encoded_input)
>>> [282, 1913, 4699, 0, 1, 282, 1913, 20127, 25, 764, 764]
```

You can verify by counting that the input sentence contains 11 tokens, counting the punctuations "." and "!" separately, matching the size of the output list of encodings. Also, notice that the last two indices are the same (i.e., 764), corresponding to the two exclamation marks "!".

Vocab also has an option to decode back the encoded input, using the lookup_tokens method.

```
decoded_input = my_vocab.lookup_tokens(encoded_input)
print(decoded_input)
>>> ['i', 'am', 'learning', '<unk>', '.', 'i', 'am', 'loving', 'it', '!', '!']
```

CHAPTER 4   INTRODUCTION TO NATURAL LANGUAGE PROCESSING: BUILDING A TEXT CLASSIFIER

Notice how "nlp" was an unknown token and hence mapped to the default index, corresponding to <unk>, by the vocab object (and not by the tokenizer). While looking up the corresponding token, i.e., while decoding, it mapped this default index to the <unk> token.

Data Loader

As before, we use a data loader to load the data into batches of input tensors. In this case, our data contains text samples, and we need to communicate to the DataLoader class that it should collate together the different samples into a batch. This is achieved using a collate function that is passed as an argument to the DataLoader.

We write our custom collate function here:

```
from torch.utils.data import DataLoader

def my_collate_fn(batch):
    input_encodings, label_list, offsets = [], [], [0]
    for label, text in batch:
        label_list.append(int(label)-1)
        text_encoding = my_vocab(tokenizer(text))
        input_encodings += text_encoding
        offsets.append(len(text_encoding) + offsets[-1])

    input_encodings = torch.tensor(input_encodings)
    offsets = torch.tensor(offsets[:-1])

    inputs = {"input_encodings": input_encodings, "offsets": offsets}
    label_list = torch.tensor(label_list, dtype=torch.int64)
    return inputs, label_list
```

Let us zoom in on what we achieve in this collate function.

1. **Labels**: The labels, as we saw before, belong to the set [1,2,3,4]. We subtract 1 from each label before appending it to a list, which is finally returned as a torch tensor. Having 0-indexed labels will help us later when we write our model definition.

2. **Text**

    a. Each text sample is passed through our tokenizer, followed by our vocab object, into a variable called `text_encoding`.

    b. It is then appended to a list called `input_encodings`. Having all our input encodings into a single long list might seem weird at first – note again that input_encodings is a single list which contains word encodings of all the words in our batch (and it is not a list of lists).

    c. How will the model be able to distinguish between different batch samples? For that, we maintain another list of `offsets` that stores the starting point of each sequence in our batch. So offsets[0] (which is always 0) contains the start index of the first sample in our batch, offsets[1] points to the starting index of the second sample in our batch, and so on.

    d. We pack these input_encodings and offsets in a dictionary together and return them from the collate function. As we shall see later, we use the nn.EmbeddingBag layer, for which it suffices to pass these input encodings along with offsets, and it can figure out the batch contents and do its job.

Note that the last iteration of the for-loop will add the total length of `input_encodings` as the last element of the list `offsets`, which we don't require. Hence, we remove this last element from `offsets` at the end.

We can test the contents of our dataloader by feeding it the train_set, specifying the batch_size of 8, and passing our collate function to it.

```
dataloader = DataLoader(
    train_set, batch_size=8, shuffle=False, collate_fn=my_collate_fn
)
```

I would encourage you to print out different characteristics of "text" and "label" within this dataloader object to understand what it exactly contains. For example, you can iterate over it and print some of its attributes:

```
for text,label in dataloader:
    print(text['input_encodings'].shape)
    print(text['offsets'])
```

First Model: Training the Embedding Layer

Our first model (see Figure 4-6) will basically pass each token in an input sequence through an Embedding layer and then take the mean of all these embeddings. After that, it will contain a fully connected hidden layer, followed by another fully connected output layer, which will give four output logits, since we have four classes. These output logits will be used in cross-entropy loss, which internally applies softmax and computes the loss. During inference, these logits can either be passed through the softmax function to obtain class probabilities, or we can simply take the class with the highest logit value as our prediction.

Here is how we code it up:

```
from torch import nn

class TextClassificationModel1(nn.Module):
    def __init__(self, vocab_size, embedding_dim, num_classes):
        super(TextClassificationModel1, self).__init__()
        self.embedding_bag = nn.EmbeddingBag(vocab_size, embedding_dim,mode='mean',max_norm=1.0)
        nn.init.uniform_(self.embedding_bag.weight, -0.5, 0.5)

        self.fully_connected1 = nn.Linear(embedding_dim, 20)
        self.fully_connected2 = nn.Linear(20, num_classes)

    def forward(self, text):
        input_encodings, offsets = text['input_encodings'], text['offsets']

        embed_mean = self.embedding_bag(input_encodings, offsets)
        fc1 = self.fully_connected1(embed_mean)
        fc2 = self.fully_connected2(fc1)
        return fc2
```

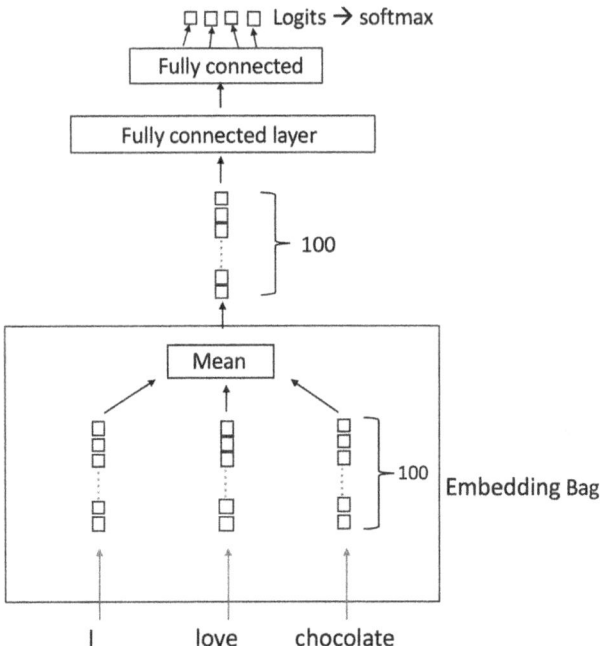

Figure 4-6. Architecture for model 1

Embedding Layer in PyTorch

Before we dive deeper into the architecture, let us see how an nn.Embedding layer works in PyTorch. We saw before that an input sequence is first encoded into a list of integer indices. An nn.Embedding layer is just a lookup table mapping these indices to vectors of a specified dimension, which is also called the embedding dimension. These mappings are typically initialized randomly and then trained during the model training phase. Alternatively, they can be initialized from a pretrained model like Word2Vec and then either trained along with the model or kept frozen while the model gets trained.

EmbeddingBag Layer

The EmbeddingBag layer performs two steps on an input sequence – embeddings followed by aggregation. It first passes all its input tokens through an embedding layer and then applies an aggregation like mean, sum, or max, over all these embeddings. This yields a single vector with a size equal to the embedding dimension.

Now, let us take a closer look at the layers of our model with the running example of the sentence: "I love chocolate":

1. Embedding Bag Layer: As described in the previous paragraph, this will take the embeddings of each of the words or tokens, i.e., "I", "love", "chocolate". We shall be choosing our embedding dimension as 100. So this would give us three vectors of dimension 100 each – these are the embedding vectors, which our model will learn during training. The embedding bag layer will then take the mean of these three vectors to get a single vector of dimension 100. The max_norm is set to 1.0, which restricts the embedding vectors to have a maximum norm of 1.0 by renormalizing them.

2. Following the Embedding bag layer, we have a hidden layer, which is a fully connected layer with `embedding_dim` inputs, i.e., 100 inputs in our case, and 20 outputs.

3. Finally, we have another fully connected layer with 20 inputs and `num_classes` outputs, i.e., 4 outputs in our case. This final layer will output four floating-point numbers, also called logits. We shall be using the CrossEntropyLoss function as the loss function, which converts these logits into probability scores using softmax and calculates the cross-entropy loss.

Also, we initialize the embedding weights uniformly between -0.5 and 0.5 by using the function `nn.init.uniform_()`.

Embedding Layer Instead of EmbeddingBag

We could also have achieved this same effect by using an nn.Embedding layer followed by torch.mean. Using nn.EmbeddingBag has two main advantages:

1. **Efficiency**: This function has been optimized so that it computes the mean (or any of the other modes like sum, max) without instantiating the intermediate embeddings.

2. **Ease**: Had we used the Embedding layer, we would need all the inputs in each batch to be of equal length, which would require us to pad the input first. Moreover, while taking the mean, we would need to consider only the non-padded inputs. EmbeddingBag allows us to avoid this hassle.

Training, Validation, and Testing

The training, validation, and testing loops are almost exactly the same as those in Chapter 3, except for a few small changes. We reproduce them here for ease, without going too much into the details.

Training

One important addition to the training loop compared to that in Chapter 3 is the clip_grad_norm_ function, which clips the gradient norms during the backward pass. This helps avoid the numerical problems that may arise if and when the gradients become too large and helps with stability.

```
def train_loop(dataloader, model, loss_fn, optimizer):
    data_size = len(dataloader.dataset)
    batch_size = dataloader.batch_size
    model.train() # Sets the model to training mode
    total_loss = 0
    print(f"Training set:")
    for batch_idx, batch_data in enumerate(dataloader):
        # Compute predictions and loss for this batch
        text, label = batch_data

        pred = model(text)
        loss = loss_fn(pred, label)
        total_loss += loss.item()

        # Backpropagation and parameter updates
        loss.backward()
        torch.nn.utils.clip_grad_norm_(model.parameters(), 0.1)
```

```
        optimizer.step()
        optimizer.zero_grad()

        # Print loss for every 100 batches
        if batch_idx % 100 == 99:
            print(f"Loss for batch {batch_idx+1:>3d}: {total_
            loss/100:>4f}")
            total_loss = 0
```

Early Stopping and Validation Loop

The early stopping and validation loops are exactly the same as in Chapter 3. We reproduce them here for reference:

```
class EarlyStopper:
    def __init__(self, min_delta = 0.001, patience = 5):
        self.patience = patience
        self.min_delta = min_delta
        self.improvement_counter = 0
        self.best_val_loss = float('inf')

    def should_stop(self, val_loss):
        if val_loss < self.best_val_loss - self.min_delta:
            # sufficient improvement
            self.improvement_counter = 0
            self.best_val_loss = val_loss
        else:
            self.improvement_counter += 1
            if self.improvement_counter >= self.patience:
                return True
        return False

def val_loop(dataloader, model, loss_fn, early_stopper):
    data_size = len(dataloader.dataset)
    batch_size = dataloader.batch_size
    model.eval() # Sets the model to eval (non-training) mode
```

```
    num_batches = len(dataloader)
    val_loss, correct_pred = 0, 0

    with torch.no_grad(): # Don't need gradients during validation
        for text, label in dataloader:
            pred = model(text) # prediction vectors
            val_loss += loss_fn(pred, label).item()
            pred_label = pred.argmax(1)
            correct_pred += (pred_label == label).long().sum().item() #
            Refer to exercise in chapter 2
    val_loss /= num_batches
    pred_accuracy = 100.0*correct_pred / data_size
    print(f"Validation set: \n Accuracy: {pred_accuracy:>0.1f}%, Avg loss:
    {val_loss:>6f} \n")
    if early_stopper.should_stop(val_loss):
        return True
    return False
```

Test Loop

The test loop is also quite similar to the one in Chapter 3.

```
def test_loop(dataloader, model, loss_fn):
    data_size = len(dataloader.dataset)
    batch_size = dataloader.batch_size
    model.eval() # Sets the model to eval (non-training) mode

    num_batches = len(dataloader)
    test_loss, num_correct_pred = 0, 0

    with torch.no_grad(): # Don't need gradients during inference
        for text, label in dataloader:
            pred = model(text)
            test_loss += loss_fn(pred, label).item()
            pred_label = pred.argmax(1)
            num_correct_pred += (pred_label == label).long().sum().item() #
            Refer to exercise in chapter 2
```

```
test_loss /= num_batches
pred_accuracy = 100.0*num_correct_pred / data_size
print(f"Test set: \n Accuracy: {pred_accuracy:>0.1f}%, Avg loss:
{test_loss:>6f} \n")
return pred_accuracy
```

Putting It All Together

We first split the train dataset further into training and validation data in a 9:1 ratio. We also define the train_dataloader, val_dataloader, and test_dataloader with our version of the collate function defined in the "Data Loader" section.

```
val_flag = True
train_validation_ratio = 0.1
batch_size = 64

if val_flag:
    train_data, val_data = random_split(train_set_map, [1 - train_
    validation_ratio, train_validation_ratio])
    val_dataloader = DataLoader(val_data, batch_size=batch_size,
    shuffle=True, collate_fn=my_collate_fn)

train_dataloader = DataLoader(train_data, batch_size=batch_size,
shuffle=True, collate_fn=my_collate_fn)
test_dataloader = DataLoader(test_set_map, batch_size=batch_size,
shuffle=False, collate_fn=my_collate_fn)
```

Finally, it is just a matter of calling the training loop along with early stopping and the learning rate scheduler. This again is quite similar to our template used in Chapter 3.

```
from torch.optim.lr_scheduler import ExponentialLR
torch.manual_seed(9) # To reduce variability in each run

num_classes = len(set([label for (label, text) in train_set_map]))
vocab_size = len(my_vocab)

# Hyperparameters
learning_rate = 2
```

```
num_epochs = 40
embedding_dim = 100

#Defining model, loss, optimizer
model = TextClassificationModel1(vocab_size, emsize, num_class).to(device)
loss_fn = nn.CrossEntropyLoss()
optimizer = torch.optim.SGD(model.parameters(), lr=learning_rate)
scheduler = ExponentialLR(optimizer, gamma=0.9)

early_stopper = EarlyStopper()
for i in range(num_epochs):
    print(f"Epoch {i+1}\n-------------------------------------------")
    train_loop(train_dataloader, model, loss_fn, optimizer)
    test_loop(test_dataloader, model, loss_fn)
    if val_loop(val_dataloader, model, loss_fn, early_stopper):
        print(f"Early stopping.")
        break
    scheduler.step()

print("Model trained!")
```

Notice that we apply the loss function CrossEntropyLoss, which expects unnormalized logits (i.e., the floating-point numbers before applying softmax) and the correct target class label as inputs. It then internally applies softmax before computing the cross-entropy loss (refer to Chapter 3 for the definition).

Results

This gave an accuracy of about 90.4% on my machine. You can try to improve upon this by a more careful hyperparameter tuning.

Second Model: Using Pretrained Word Embeddings Using Word2Vec

Our first model involved an EmbeddingBag layer, which computed the mean of the embeddings of all the tokens in each input sample, and then passed this mean to a hidden layer, followed by an output layer, which then predicted the class. This EmbeddingBag layer was trained as part of the model training process.

However, there are several models available as part of libraries that have been pretrained on large volumes of data to produce word embeddings. In our second model, we shall use one such model called Word2Vec, which has been trained on the Google News dataset. Although such models may not have been trained exactly on our dataset, they are usually trained on a diverse enough set of considerable volume, so that they may end up giving useful results for our task at hand.

This second model will consist of a pretrained Word2Vec model, which maps each token to a 300-dimensional embedding vector. The model then takes the mean of all the embedding vectors of all the tokens in the input sample. Post this embedding layer, the architecture mimics that of model 1, i.e., a hidden layer followed by an output layer predicting the class.

The Word2Vec Model (Optional)

Word2Vec is a family of algorithms introduced in the paper "Efficient Estimation of Word Representations in Vector Space" by Mikolov et al. These algorithms aim to build continuous vector representations of words from a given dataset. The main characteristic that we expect from these representations is that words that are similar to each other in meaning should have their corresponding vectors near each other in the vector space.

One interesting observation is that these representations often are good at analogies, as implemented by the + and – operations. An example that is ubiquitous in the NLP literature is that if you start with the vector representing the word "King", v_{King}, and subtract from it the vector representing "man", v_{man}, while adding the vector representing the word "woman", v_{woman}, then the resultant vector will be nearest to the vector representing the word queen. In short, $v_{King} - v_{man} + v_{woman}$ is nearest to the representation v_{Queen}.

Another specialty of the Word2Vec algorithm is that it is an unsupervised learning algorithm, i.e., it does not require labelled data for training. It is able to do so using techniques like masking each word one at a time, trying to predict it from the surrounding words, and other similar approaches. The details of these algorithms are not relevant to our discussion. But what is noteworthy is that not requiring labelled data allows these models to train on huge volumes of data.

CHAPTER 4   INTRODUCTION TO NATURAL LANGUAGE PROCESSING: BUILDING A TEXT CLASSIFIER

Loading the Word2Vec Model

You can download the pretrained Word2Vec model trained on Google News as follows:

```
import gensim.downloader as api
w2v = api.load('word2vec-google-news-300')
```

This model, which we can now access as w2v, maps each word in its vocabulary to a 300-dimensional embedded vector. We can then access the embedded vector corresponding to a word by simply indexing as w2v[word].

```
word1 = "Success"
word2 = "Thisisnotaword"

print(w2v[word1])
print(w2v[word2])
```

You will see that this first prints the embedded 1D vector of length 300 corresponding to the word word1. However, if a word is not found in its vocabulary, it throws a keyError exception; for this reason, we would always have to check if a word is present in the vocabulary before trying to access its embedded vector.

We can check the size of its vocabulary as

```
vocab_size = len(w2v.index_to_key)
print(f'The vocab size of Word2Vec is: {vocab_size}')
>>> The vocab size of Word2Vec is: 3000000
```

You can also try printing five random words from the vocabulary of your pretrained Word2Vec model[2]:

```
for i in range(5):
    rand_idx = random.randint(0,vocab_size-1)
    print(f'The word at randomly chosen index {rand_idx} is {w2v.index_to_key[rand_idx]}')
```

Check out this tutorial (https://radimrehurek.com/gensim/auto_examples/tutorials/run_word2vec.html) on Word2Vec to learn more about its functionalities.

---

[2] Omitting the output of this code since it is very machine-specific. Check it out on your own machine!

Collate Function

The first thing we need to do differently for this model is the collate function. Since the w2v embedding layer does not need training, we include it in the collate function itself. In essence, the collate function takes the w2v embeddings of all the tokens in the given input sample, calculates their mean, and appends it to be returned as a tensor containing one 300-dimensional vector per input sample. This vector can be thought of as a succinct numerical representation of the entire input sequence.

```
def collate_fn_word2vec(batch):
    text_list, label_list = [], []

    for label, text in batch:
        label_list.append(int(label)-1)
        text_tokens = tokenizer(text)

        # Filter tokens belonging to the w2v vocab
        text_tokens_in_w2v = [token for token in text_tokens if
        token in w2v]

        # Create a tensor of token embeddings and take the mean
        w2v_embeddings = torch.tensor([w2v[token] for token in text_tokens_
        in_w2v])
        w2v_embed_mean = torch.mean(w2v_embeddings, dim=0) if len(text_
        tokens_in_w2v) > 0 else torch.zeros_like(torch.tensor(w2v[0]))
        text_list.append(w2v_embed_mean)

    text_list = torch.stack(text_list,dim=0)
    label_list = torch.tensor(label_list, dtype=torch.int64)
    return text_list, label_list
```

This requires that the Word2Vec model has been loaded in the variable w2v. The tokens not belonging to the vocabulary of Word2Vec are ignored, i.e., they don't contribute to the embedding mean. In the extreme case when none of the tokens of a sample are found in the vocabulary, the mean would simply be the all-zero tensor.

Note that each of the embedding mean tensors is 300-dimensional, and therefore, we don't need any padding or offset information. The torch.stack function is then used to concatenate all these embedding mean tensors into a 2D tensor of dimension [batch_size, embedding_dim].

Model Architecture

As we saw, the embeddings were already computed by the collate function. The model input now consists of a batch with a 300-dimensional tensor for each sample. The architecture simply contains a hidden layer of size 20, followed by an output layer that outputs four logits.

```
class TextClassificationModel2(nn.Module):
    def __init__(self, vocab_size, embedding_dim, num_classes):
        super(TextClassificationModel2, self).__init__()
        self.fully_connected1 = nn.Linear(embedding_dim, 20)
        self.fully_connected2 = nn.Linear(20, num_classes)

    def forward(self, text):
        fc1 = self.fully_connected1(text)
        fc2 = self.fully_connected2(fc1)
        return fc2
```

All the other parts of the code remain unchanged. We simply change the embedding dimension to match that of the Word2Vec vectors (i.e., 300). Also, we would have to pass our new collate function `collate_fn_word2vec` to the DataLoader constructor.

Results

Running this model on my machine gives an accuracy of 88.8%.

Alternative: Training the Word2Vec Model from Scratch

The model we loaded, i.e., Word2Vec, was trained on Google News, which was a very similar corpus to our data. Had our dataset been significantly different qualitatively, we may also have considered training the Word2Vec model from scratch on our data. The dataset needs to be large enough for this to be a viable strategy.

We now briefly explore how to train the Word2Vec model on our training data from scratch and use it to create word embeddings. These embeddings can then be used exactly like in model 2.

Training Word2Vec from Scratch

We can pass our training data to train a Word2Vec model as follows:

import gensim

```
token_list = [tokenizer(text) for label,text in train_iter]
w2v_embedding_model = gensim.models.Word2Vec(token_list, min_
count=1,vector_size=100, window=5)
```

We create token_list, which is a list of lists, where each inner list contains all the tokens belonging to an input text sample. This token_list is passed as an input to genism.models.Word2Vec, along with the following other training parameters:

1. **min_count**: The minimum number of times a word has to occur in the data to be included in the vocabulary

2. **vector_size**: Representing the dimension of the embedding vectors

3. **window**: Representing the window size used in the Word2Vec algorithm

The Word2Vec model can be accessed as w2v_embedding_model.wv:

```
print(f'Vocabulary size of our trained Word2Vec model: {len(w2v_embedding_
model.wv.index_to_key)}')
```

The behavior of w2v_embedding_model.wv is the same as the variable w2v in model 2, and therefore, this can be used directly instead of w2v in the collate function, with the same model architecture as model 2.

Text Classification with Pretrained Hugging Face Transformers

In model 2, we used Word2Vec models to obtain word embeddings, which were then passed to a dense layer for classification. Word2Vec embeddings capture the syntactic and semantic similarities of words, but often miss the context-dependent meaning of a word. For example, it may not be able to capture the differences between the word "bank" used in the context of a "river bank" vs. the "bank of America".

Transformers, discussed in "The Transformer Architecture" section, are much more proficient at capturing the contextualized meaning of a word. BERT (Bidirectional Encoder Representations from Transformers) is a popular transformer-based encoder architecture that considers bidirectional context while creating its representations.

Our approach in this section will be to simply replace Word2Vec embeddings with these superior BERT-based encoder vectors and pass them to a neural network-based classification head, which we shall train.

BERT (Bidirectional Encoder Representations from Transformers)

BERT with HuggingFace: A Quickstart Tutorial

Hugging Face Hub is a host containing many latest NLP and computer vision models that are pretrained by different research groups and big companies. The next chapter is dedicated to practical NLP techniques using Hugging Face models. In this section, we shall use Hugging Face to load DistilBERT, which is a faster, lightweight variant of BERT.

Tokenizer

We will load the tokenizer for the DistilBERT base model using the AutoTokenizer class.

```
from transformers import AutoTokenizer

model_name = "distilbert-base-uncased"
tokenizer = AutoTokenizer.from_pretrained(model_name)
```

We now use the tokenizer on the following sample sentences using some specific argument values, which we shall soon explain:

```
text1 = "I love chocolate."
text2 = "The quick brown fox jumps over the lazy dog"

tokenizer([text1,text2], add_special_tokens=True, padding='longest',truncation=True,return_tensors='pt')
>>> {'input_ids': tensor([[ 101, 1045, 2293, 7967, 2015, 1012,  102,
    0,    0,    0,
             0],
```

```
[  101, 1996, 4248, 2829, 4419, 14523, 2058, 1996,
13971, 3899,
    102]]), 'attention_mask': tensor([[1, 1, 1, 1, 1, 1, 1, 0,
    0, 0, 0],
[1, 1, 1, 1, 1, 1, 1, 1, 1, 1, 1]])}
```

Let us see what the different arguments mean:

1. We set "add_special_tokens" to True. To see what this exactly does:

    ```
    tokenizer.tokenize(text1, add_special_tokens=True)
    >>> ['[CLS]', 'i', 'love', 'chocolates', '.', '[SEP]']
    ```

    Special tokens [CLS] and [SEP] were added, where [CLS] is added at the beginning of each input, whereas [SEP] is a separator token separating two inputs or separating question and answer in a question-answering setting. The [CLS] token has special significance because the model is trained to aggregate the essence of the entire sequence in its [CLS] token.

2. We set padding to "longest." As can be seen, this pads the first sentence to match the length of the longer second sentence. In general, each sequence is padded to match the length of the longest sequence in a batch.

3. We set truncation to True, which in this context means that any sequence will be truncated to the max length that is allowed by the pretrained model.

4. Setting return_tensors to "pt" returns PyTorch tensors in the output. You can also choose other options like TensorFlow tensors, etc.

Calling the tokenizer outputs a dictionary with input_ids and attention_mask as the keys. The input_ids contains the tokenized versions of inputs after padding and truncation. The attention mask contains 1s at positions that correspond to input tokens and 0s wherever padding was added.

Notice that the 101 at the beginning of each of the encodings must represent [CLS], whereas the 102 at the end represents [SEP].

DistilBERT Model

We now load the DistilBERT base pretrained model using the AutoModel class.

```
from transformers import AutoModel

model = AutoModel.from_pretrained("distilbert-base-uncased")
out = model(input_ids=tokens['input_ids'],
attention_mask=tokens['attention_mask'])
print(out.last_hidden_state.shape)
>>> torch.Size([2, 11, 768])
```

The model output contains a last_hidden_state attribute, which contains the dense vector representation of our two input sequences.

We gave as inputs the input_ids and attention_mask from the tokens dictionary, which we had computed using the tokenizer in the previous subsection.

Notice the shape of last_hidden_state: the 0th dimension has size 2 since the input had two samples (text1 and text2). After padding, the tokens generated by the tokenizer contained 11 input_ids in each input sequence; hence, the first dimension is 11. Finally, the second dimension is 768 since this model is configured to produce 768-dimensional vector encodings as hidden states.

In summary, this model outputs a dense vector representation of dimension 768 for every token in every input sample, including the special tokens.

Our approach using DistilBERT for text classification would be to use these dense representations just like we used embeddings from the Word2Vec model in the previous section. In this case, as we saw before, the model stores the essence of the entire sequence, or in other words, an aggregation of the sequence, in the dense representation of the special token [CLS]. So we will just be passing the vector corresponding to [CLS] to the fully connected layers as before and train this classifier.

Using DistilBERT for Text Classification

The main changes would be to the collate function and the model architecture.

Collate Function

```
from transformers import AutoModel, AutoTokenizer

model_name = "distilbert-base-uncased"
tokenizer = AutoTokenizer.from_pretrained(model_name)

def collate_fn_distilbert(batch):
    text_list, label_list = [], []
    for label, text in batch:
        label_list.append(int(label)-1)
        text_list.append(text)

    tokens = tokenizer(text_list, add_special_tokens=True,
    padding='longest',truncation=True,return_tensors='pt')

    label_list = torch.tensor(label_list, dtype=torch.int64)
    return tokens, label_list
```

As before, we load the tokenizer for distilbert-base-uncased and use it to tokenize the entire text_list in the batch. We use the same config arguments to the tokenizer as used in the previous subsection.

Model Architecture

We shall now define our model architecture:

```
class TextClassificationModel3(nn.Module):
    def __init__(self, embedding_dim, num_classes):
        super(TextClassificationModel3, self).__init__()

        self.encoder_model = AutoModel.from_pretrained(model_name)

        for param in self.encoder_model.parameters():
            param.requires_grad = False

        self.fully_connected1 = nn.Linear(embedding_dim, 50)
        self.fully_connected2 = nn.Linear(50, num_classes)
```

```
def forward(self, input_ids, attention_mask):
    encodings = self.encoder_model(input_ids, attention_mask).last_
    hidden_state[:,0,:]

    fc1 = self.fully_connected1(encodings)
    fc2 = self.fully_connected2(fc1)
    return fc2
```

We load the DistilBERT model as self.encoder_model. We freeze its parameters since we don't wish to train the DistilBERT model further. Then we use a hidden layer of size 50, followed by an output layer for classification, as in the previous models. We use a higher hidden state dimension of 50 (as opposed to 20 in models 1 and 2) simply because we are starting with embedding vectors of much higher dimensions here, i.e., 768 as opposed to 100 and 300 in models 1 and 2, respectively.

Flexibility

Notice how adaptable this approach is. We can simply change the model name to load a whole different model from Hugging Face for the same task with no (or minimal) additional changes. Similarly, we could also fine-tune the DistilBERT model by not freezing its parameters.

Results

Training takes significantly longer for this model since DistilBERT is a much bigger model, which takes some significant time to generate the embeddings themselves. Even so, it reached an accuracy of 88.5% after just one epoch of training on my machine.

Summary

- Natural language processing (NLP) involves comprehending, processing, and analyzing human languages.

- It focuses on solving diverse problems like translation, summarization, dialogue systems, text classification, parts of speech tagging, question answering, etc.

- Raw text input usually needs to be preprocessed by first tokenizing each word (or sometimes characters or subwords) into separate tokens and then mapping each word in the vocabulary to a distinct integer index.

- A typical NLP model then uses words to convert these token indices into dense vector representations. These representations can then be passed through a classifier for text classification or processed in other suitable ways for different tasks.

- This process of converting text into dense vector representations is achieved by the encoder architecture, which can be implemented by RNNs, LSTMs, transformers, etc.

- The dense vector representations given by an encoder can then be converted back into text in a format specific to the task by an architecture called the decoder. This can again be implemented via RNNs, transformers, etc.

- Encoder-decoder architecture can convert the input text into the desired output text based on the task at hand. The models achieving this are generically called sequence-to-sequence models.

- We worked on a text classification project of classifying news samples into one of four fixed categories using the AGNews dataset.

- Our first approach involved training an embedding bag layer – which takes the mean of embeddings of all the words in an input sequence – followed by a hidden layer and an output classification layer.

- In our second approach, we used the pretrained Word2Vec model to create word embeddings, which were then passed to a classifier that we trained on our data.

- We then saw a quick tutorial for using Hugging Face to load the pretrained DistilBERT transformer model.

- We used the DistilBERT model to produce dense vector representations of our input sequences. We passed these representations as input to our classifier, which we then trained.

CHAPTER 5

Practical Natural Language Processing with Hugging Face

Many of the world's leading tech companies and research groups now make their state-of-the-art pretrained models publicly available on the Hugging Face Hub. Whether you want to build a quick AI app or are just getting started with your dream AI project, it is incredibly worthwhile to plug one of these models into your use case and see how it performs. This may solve your problem right away or at least provide a solid foundation that you can refine later. At the very least, it will establish a reasonable baseline for your future experiments. For this reason, becoming well-acquainted with the techniques for quickly loading and utilizing these models is an essential skill for any modern ML practitioner.

This chapter will guide you through several key NLP tasks by leveraging the power of pretrained models from the Hugging Face ecosystem. We will begin in the "Hugging Face" section with an introduction to the key components of Hugging Face. From there, we will embark on a series of practical mini-projects, starting in the "Text Classification" section with **text classification**. We will first get a feel for the task with a quick zero-shot classification example before diving into the more involved process of fine-tuning a model for high accuracy. In the "Summarization" section, we will tackle **summarization**, where we will first summarize a Wikipedia article using a pretrained model, further improving it with a chunking approach. We will follow it up with another project where we will fine-tune a pretrained model for summarizing some informal chat conversations. Finally, in the "Question Answering" section, we will explore **question answering**, demonstrating how to use a pretrained model to find precise answers within a text.

Hugging Face
What Is Hugging Face

Hugging Face is an online platform that enables collaborative effort in machine learning, more specifically focused on NLP. It is a space where ML practitioners can upload datasets and pretrained models for others to use. It is best understood as an **ecosystem** consisting of two main parts:

1. **The Hub:** An online platform for sharing and collaborating on models, datasets, and ML applications

2. **Open Source Libraries:** A suite of libraries like `transformers`, `datasets`, and `evaluate` that make it incredibly easy to download and use these tools in your own code

Hugging Face thrives on a collaborative spirit. It is incredibly user-friendly, where anyone, from a novice training their first model to an ML expert, is encouraged to share their work.

The Model Hub

The Model Hub is a vast repository of thousands of pretrained models. You can explore them here: `https://huggingface.co/models`.

The left pane (Figure 5-1) contains different categories for filtering models based on tasks. It has categories for multimodal, computer vision, natural language processing, audio, tabular, etc. Within each category, it has several subcategories, for example, NLP has text classification, summarization, question answering, and many more. This makes it easy to find a high-performing, pretrained starting point for almost any project you can imagine.

Figure 5-1. *The left pane with task-based filters for models on the Hugging Face Hub*

Hugging Face is especially famous for its powerful transformer library.

The Dataset Hub

Hugging Face also hosts several datasets: `https://huggingface.co/datasets`. It has similar filters to Hugging Face models, based on tasks, etc. The datasets library allows you to load any of these with a single line of code, handling all the downloading, caching, and processing for you.

The Pipeline API: Instant Inference

The transformers library in Hugging Face has an abstraction called pipeline that offers a simple API to load and use different types of models directly for inferencing. As you will soon see in our upcoming projects, for example, a summarization model can be loaded as simply as

```
summarizer = pipeline("summarization", model="facebook/bart-large-cnn")
```

Even the "model" parameter can be skipped, in which case it will default to some (predecided) model that can perform the summarization task.

Fine-Tuning: Specializing Pretrained Models

The most commonly used workflow in modern NLP is **fine-tuning**. This involves taking a pretrained model and training it further on your specific dataset to adapt it to your unique task. The transformers library provides two main paths for this:

1. **The Trainer API:** This is a high-level training API that handles the entire training process for you. It automates device placement (using a GPU if available), batching, evaluation, and checkpointing. **This is the recommended approach for most applications and is what we use in this book.**

2. **Native PyTorch/TensorFlow:** For researchers or advanced users who need maximum control over the training process, it is possible to write a custom training loop in native PyTorch or TensorFlow while still using Hugging Face models and utilities.

The standard fine-tuning workflow, which we've followed in our projects, consists of the following steps:

1. **Load and Prepare the Data:** Use the datasets library to load a dataset and an AutoTokenizer to tokenize the text.

2. **Load the Pretrained Model:** Use an AutoModelFor... class (like AutoModelForSequenceClassification) to load a base model with a new, untrained head for your specific task.

3. **Set Up the Training Process:** Configure the `TrainingArguments` to control all hyperparameters, set up a `DataCollator` for intelligent batching, and define a `compute_metrics` function for evaluation.

4. **Train the Model:** Instantiate the `Trainer` object with all the components from the previous steps, and call the `.train()` method to launch the fine-tuning process.

5. **Evaluation and Inference:** After training, use `.evaluate()` or `.predict()` to measure the model's performance on a held-out test set, and use the `pipeline()` for easy inference on new data.

For a more detailed explanation, here is an excellent tutorial on fine-tuning Hugging Face models: `https://huggingface.co/docs/transformers/en/training`.

Text Classification

Project 1: Zero-Shot Classification

Let's say you are given a single sentence, and you want to classify it into one of a fixed set of categories. But you don't have training data for these types of sentences with those exact categories. This is the problem addressed by zero-shot classification. It arranges to use pretrained models to classify any given sentence into any given set of labels. Note that the labels are not fixed but specified as an input as well!

This approach is incredibly useful for rapid prototyping, establishing a quick baseline, or for situations where you don't have enough data to train a custom model.

In our mini-project, we will see how well a general-purpose, zero-shot model can classify emotions, given a short input text. We will follow up with a more detailed project on emotion classification via fine-tuning a pretrained model.

The transformers library makes this advanced technique accessible through its pipeline API. We'll use a model specifically trained on natural language inference (NLI), which is cleverly repurposed to handle zero-shot classification.

```
from transformers import pipeline

# 1. Load the zero-shot-classification pipeline specifying the model
classifier = pipeline("zero-shot-classification",
                     model="facebook/bart-large-mnli")
```

```
# 2. Define some text sequences we want to classify
sequence_to_classify_1 = "I am so excited for this concert tonight, I've
been waiting for it for months!"
sequence_to_classify_2 = "I'm not sure about this new project. The deadline
is so tight and I feel anxious."

sequences = [sequence_to_classify_1, sequence_to_classify_2]

# 3. Define the candidate labels
candidate_labels = ['sadness', 'joy', 'love', 'anger', 'fear', 'surprise']

# 4. Run the classification
results = classifier(sequences, candidate_labels)

# 5. Print the results in a readable format
for text, result in zip(sequences, results):
    print(f"Text: \"{text}\"")
    print(f"  -> Predicted Emotions: {result['labels']}")
    print(f"  -> Confidence Scores: {[round(s, 4) for s in
    result['scores']]}\n")

Text: "I am so excited for this concert tonight, I've been waiting for it
for months!"
  -> Predicted Emotions: ['joy', 'surprise', 'love', 'anger', 'fear',
    'sadness']
  -> Confidence Scores: [0.6266, 0.2849, 0.0608, 0.0146, 0.0085, 0.0047]

Text: "I'm not sure about this new project. The deadline is so tight and I
feel anxious."
  -> Predicted Emotions: ['fear', 'surprise', 'sadness', 'anger',
    'love', 'joy']
  -> Confidence Scores: [0.4943, 0.4081, 0.0474, 0.0366, 0.0078, 0.0058]
```

To get our classifier, we only needed to call the pipeline() function, specifying the task as "zero-shot classification" and the underlying model as "facebook/bart-large-mnli". Similarly, there are many other models that have been trained on NLI data, which could also be used for zero-shot classification for text.

As you can see, the model correctly predicted "joy" as the label for the first text and "fear" for the second. However, it did so with relatively low confidence scores, especially for the second sentence. It did so even though the emotion of the second sentence was abundantly clear.

Project 2: Fine-Tuning Pretrained Models for Emotion Classification

We now move on to one of the most commonly used techniques in practical NLP: fine-tuning a pretrained model. Large technology companies and research labs have made powerful language models publicly available. These models, having been trained on vast amounts of text data, possess a deep understanding of grammatical structure and the semantic relationships between words. Our task is to leverage this existing expertise of such a model and steer it toward our specific goal.

In this project, we will work with the emotions dataset, which contains Twitter messages in English containing one of the six emotions: "sadness", "joy", "love", "anger", "fear", and "surprise". We will load a pretrained model named **DistilBERT**, complete with its learned weights. We will then fine-tune it on our training data, i.e., we shall pass our data through the model and use backpropagation to update its weights. Finally, we will evaluate this fine-tuned model using held-out test data. Held-out data refers to a portion of the dataset set aside for evaluation that is not used during model training.

Imports

You may need to first install the following package.

```
!pip install datasets==3.6.0
```

Here are some imports we would need throughout this project.

```
import torch
from torch.utils.data import Dataset
import matplotlib.pyplot as plt
from torch import nn
from torch.utils.data import random_split
import matplotlib.pyplot as plt
import time
from transformers import AutoTokenizer
```

Creating a Hugging Face Account (Optional)

While not strictly required for our projects, creating a free Hugging Face account and logging in from your notebook is highly recommended. Authenticating your session allows you to save your fine-tuned models directly to your personal profile on the Hub with a single command from the notebook. This is useful for building a portfolio of your work, sharing your models with the community, or directly loading them into other applications.

To log in, you can run the following code, which will ask for an access token that can be generated from the Hugging Face portal.

```
from huggingface_hub import notebook_login
notebook_login()
```

Loading the Dataset

As we saw before, Hugging Face hosts many different datasets that can be loaded easily using the load_dataset function. In this project, we shall be working with the emotion dataset.

```
from datasets import load_dataset

emotion_dataset = load_dataset("emotion")
print(emotion_dataset)

>>> DatasetDict({
    train: Dataset({
        features: ['text', 'label'],
        num_rows: 16000
    })
    validation: Dataset({
        features: ['text', 'label'],
        num_rows: 2000
    })
    test: Dataset({
        features: ['text', 'label'],
        num_rows: 2000
    })
})
```

This gives a DatasetDict object, which is a dictionary where the keys are the splits, i.e., "train", "validation", and "test". The value for each key is the Dataset object containing the data corresponding to each split.

The Hugging Face Dataset

Let us take this opportunity to learn more about the Hugging Face Dataset class with the help of this running example.

Size

You can gauge the number of rows of a Dataset by accessing its attribute "num_rows". In this case, the `num_rows` variable is printed as part of the Dataset object, indicating that the training data here contains 16000 samples, whereas the validation and test data contain 2000 samples each.

Features

Another important attribute of a Dataset is the "features" attribute. You can print it out and see for yourself:

```
print(emotion_dataset['train'].features)
{'text': Value(dtype='string', id=None), 'label':
ClassLabel(names=['sadness', 'joy', 'love', 'anger', 'fear', 'surprise'],
id=None)}
```

As you may have guessed, this contains all the columns of the Dataset. It has two columns in this case. The first is "text", which has type Value, which is essentially a way of saying that these columns contain certain simple feature values of a single data type. In this case, these values are of type string. The second column is "label", which has type ClassLabel, which defines the set of classes that all our samples belong to. These have names "sadness," "joy," etc., and are stored as integers within the dataset.

This information about the features columns was also summarized when we simply printed emotion_dataset before.

Indexing

You can access samples from the dataset by indexing using row numbers, simply as

```
print(emotion_dataset["train"][0])
>>> {'text': 'i didnt feel humiliated', 'label': 0}
```

To select multiple rows, you can use slicing similar to how it works in Python lists or PyTorch tensors. For example, `emotion_dataset['train'][0:2]` would return a dictionary containing the first two examples as lists for each column, i.e., the "text" key would map to the two examples as a list and the "label" key would map to their corresponding labels.

Functionalities

There are several useful in-built functions available for Dataset objects. We shall look at some important ones along with a short summary of their usages; you can understand them better when you see them in action in our different projects from this chapter. You can also learn more about them here.

1. **Sort**: Sorts column values for the specified column.
2. **Shuffle**: Randomly shuffles the dataset rows.
3. **Select**: Selects rows corresponding to the specified indices.
4. **Filter**: Selects rows that match the specified predicate function.
5. **Map**: Applies the given processing function to each example in the dataset, either individually or in batches. This is perhaps the most powerful function in this list, commonly used for complex transformations like tokenization. When used with `batched=True`, it leverages highly efficient back-end libraries to process thousands of examples at once.

A key feature of these methods is that they are not "in-place" operations. When you use `.map()`, `.filter()`, or `.sort()`, the library returns a **new, modified** Dataset **object**. Your original dataset (`emotion_dataset` in this case) remains completely unchanged. This is a crucial feature for ensuring reproducibility and preventing accidental data corruption during your experiments.

I would recommend playing around with different attributes and functionalities of the Dataset class to get a firm grip of the data structure you are working with.

Exploratory Data Analysis (EDA)

In keeping with our practice of understanding the data better by looking at its different statistical properties, we shall again print the frequencies of different class labels here:

```
# Extract the list of label names from the dataset features
label_list = emotion_dataset['train'].features['label'].names
print(f"List of labels: {label_list}")

# Mapping label integers to emotion
label_tensor = torch.tensor(emotion_dataset["train"]['label'])

# Frequency of each label
unique_labels, counts = torch.unique(label_tensor, return_counts=True)
print(unique_labels)
print(counts)

>>> tensor([0, 1, 2, 3, 4, 5])
>>> tensor([4666, 5362, 1304, 2159, 1937,  572])
```

The six labels are stored as 0-indexed integers, i.e., from 0 to 5. These labels correspond to the emotions of "sadness", "joy", "love", "anger", "fear", and "surprise", in that order. You can learn more details about the emotions dataset from the documentation here.

For a visual summary, we also plot the counts histogram:

```
# Plotting the histogram for label distribution in training data
plt.bar(unique_labels, counts,tick_label=label_list)
plt.xlabel("Label")
plt.ylabel("Count")
plt.title("Label Distribution in training data")

plt.show()
```

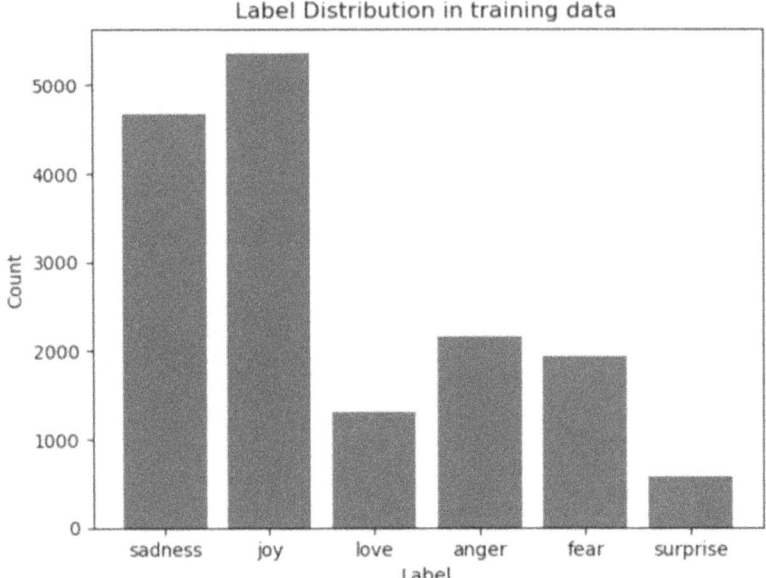

Figure 5-2. *Emotions dataset label counts*

As Figure 5-2 clearly shows, our dataset is not perfectly balanced. The "joy" and "sadness" classes have significantly more examples than "love" or "surprise". We would typically employ advanced techniques like class weighting or oversampling to help the model pay more attention to these minority classes. However, since our primary goal here is to learn about the core fine-tuning workflow in Hugging Face, we will proceed with this natural distribution to establish a strong baseline performance. We shall work with some of these advanced techniques for handling data imbalance in a later chapter (Chapter 6).

Let us also print one sample from each class to get a sense of how the sentences look like:

```
for i, label_name in enumerate(label_list):
    example = emotion_dataset["train"].filter(lambda example:
    example["label"] == i)[0]
    print(f"An example with label '{label_name}' ({i}):")
    print(f">> \"{example['text']}\"")
```

```
An example with label 'sadness' (0):
>> "i didnt feel humiliated"
An example with label 'joy' (1):
```

```
>> "i have been with petronas for years i feel that petronas has performed
well and made a huge profit"
An example with label 'love' (2):
>> "i am ever feeling nostalgic about the fireplace i will know that it is
still on the property"
An example with label 'anger' (3):
>> "im grabbing a minute to post i feel greedy wrong"
An example with label 'fear' (4):
>> "i feel as confused about life as a teenager or as jaded as a year
old man"
An example with label 'surprise' (5):
>> "ive been taking or milligrams or times recommended amount and ive
fallen asleep a lot faster but i also feel like so funny"
```

Notice the usage of the filter function to select all the rows with the specified label.

Tokenization

Before we can train a model, we must convert our text into a numerical format it can understand, via the process of tokenization (as we saw in Chapter 4). We will use a tokenizer that corresponds to the model we intend to fine-tune, which is the "distilbert-base-uncased" model. This ensures that the data preparation is consistent with the way the training data was processed for training the base (pretrained) model.

```
model_name = "distilbert/distilbert-base-uncased"
tokenizer = AutoTokenizer.from_pretrained(model_name)
```

Here, the `AutoTokenizer` function automatically downloads the correct vocabulary and tokenization rules for the specified model. We will use the `map` function of the Dataset object to apply the tokenizer to all the samples. For that reason, we will wrap the tokenizer inside a preprocess function that matches the expected format of the input to map. Map takes as input a function that expects a batch of examples as input.

```
def preprocess(examples):
    return tokenizer(examples["text"], truncation=True, add_special_
tokens=True)
```

Chapter 5 Practical Natural Language Processing with Hugging Face

We then pass this to the map function, along with a batch_size of 500:

```
emotions_tokenized = emotion_dataset.map(preprocess, batched=True, batch_size=500)
```

Let us print out an arbitrary sample from this tokenized dataset to see what it contains.

```
print(emotions_tokenized["train"][2])
{'text': 'im grabbing a minute to post i feel greedy wrong', 'label': 3, 'input_ids': [101, 10047, 9775, 1037, 3371, 2000, 2695, 1045, 2514, 20505, 3308, 102], 'attention_mask': [1, 1, 1, 1, 1, 1, 1, 1, 1, 1, 1, 1]}
```

As you can see, this contains the original text along with the label and, additionally, a list of input ids which are numerical tokens representing the words in the text, along with an attention mask which specifies which of the tokens are meaningful. Since the attention mask contains all 1s, all the tokens are important here. The attention mask uses 0s to indicate any padding tokens used during batching, telling the model to ignore them during processing.

Evaluation Metric

For this classification task, our primary evaluation metric will be **accuracy**, which measures the percentage of predictions our model gets right. We will need to wrap it in the appropriate format that can be passed to the Trainer.

```
import evaluate
accuracy = evaluate.load("accuracy")
```

The evaluate library of Hugging Face contains implementations of many of the standard evaluation metrics. We just loaded the accuracy metric here.

```
def compute_metrics(eval_predictions):
    prediction_scores, labels = eval_predictions.predictions, eval_predictions.label_ids
    predictions = np.argmax(prediction_scores, axis=1)
    return accuracy.compute(predictions=predictions, references=labels)
```

We defined the compute_metrics function to be passed to the Trainer. The argument to this is an object of type EvalPrediction, which contains attributes `predictions`, containing the model predictions, and `label_ids`, which contain the label id values.

A crucial step here is np.argmax(prediction_scores, axis=1). The model doesn't output a single label, but rather a score (or "logit") for each of the six possible emotions. The argmax function finds the index of the highest score for each example, effectively converting the raw scores into our final predicted class label, which is then compared to the reference label for accuracy computation.

Model Training

With our data prepared, we are now ready to set up and execute the model fine-tuning process. This involves three main steps: loading a pretrained sequence classification model, defining our training configuration, and then using the Trainer API to train the model.

First, we define two dictionaries, one that maps each label index to the label string and another one that does just the opposite. This is a crucial step for both configuring the model and interpreting its outputs later.

```
id2label = {i: label_list[i] for i in range(len(label_list))}
label2id = {label_list[i]: i for i in range(len(label_list))}
```

We now load the distilbert base model, captured in the model_name variable defined before, from Hugging Face. We use the AutoModelForSequenceClassification class, which automatically downloads the pretrained weights and adds a new, randomly initialized classification head on top.

```
from transformers import AutoModelForSequenceClassification, TrainingArguments, Trainer

model = AutoModelForSequenceClassification.from_pretrained(
    model_name, num_labels=6, id2label=id2label, label2id=label2id
)
```

By passing `num_labels=6` along with our `id2label` and `label2id` mappings, we configure the model's final layer to output scores for our six emotion classes.

Before we can begin training the model, we need a DataCollator. This helper function takes our tokenized examples and pads them as required to form uniform batches, which is necessary for the model to process them efficiently.

```
from transformers import DataCollatorWithPadding

# A data collator that will dynamically pad the inputs received.
data_collator = DataCollatorWithPadding(tokenizer=tokenizer)
```

We then define the training arguments to be passed to the Trainer.

```
training_args = TrainingArguments(
    output_dir="my_text_classification_model",
    learning_rate=2e-5,
    per_device_train_batch_size=32,
    per_device_eval_batch_size=32,
    num_train_epochs=5,
    weight_decay=0.1,
    eval_strategy="epoch",
    save_strategy="epoch",
    load_best_model_at_end=True,
    logging_steps=50,
    save_steps=100,
    report_to="none",
)
```

Finally, we instantiate the Trainer, which brings all our components together: the model, training arguments, datasets, data collator, and our custom metrics function. The trainer.train() command then orchestrates the entire training loop, handling device placement (using a GPU if available), batching, forward pass, backpropagation, and evaluation automatically.

```
trainer = Trainer(
    model=model,
    args=training_args,
    train_dataset=emotions_tokenized["train"],
    eval_dataset=emotions_tokenized["validation"],
```

```
        data_collator=data_collator,
        compute_metrics=compute_metrics,
)
trainer.train()
```

Let us see what some of the key training arguments mean:

1. output_dir is the directory where we want our model checkpoints and runs to be stored.

2. learning_rate and num_train_epochs are hyperparameters related to model training.

3. Per_device_train_batch_size and per_device_eval_batch_size, as the names suggest, define the batch size to be used per GPU or CPU for training and evaluation, respectively.

4. Weight_decay is the regularization multiplier, which helps the model avoid overfitting.

5. eval_strategy and save_strategy define when the Trainer should run an evaluation and save a model checkpoint, respectively. In our case, the Trainer will execute both these actions at the end of every epoch.

6. load_best_model_at_end ensures that after training is complete, the trainer.model object will be the one with the best performance on the validation set, not necessarily the one from the final epoch.

Model Testing

With our model fully trained, the final step is to evaluate its performance on the held-out test set. This provides an unbiased assessment of how well our fine-tuned model generalizes to new data it has never seen before.

This can be achieved by simply using the Trainer's evaluate function, passing the tokenized test data as input.

```
test_metrics = trainer.evaluate(emotions_tokenized["test"])
print(test_metrics)
```

```
>>> {'eval_loss': 0.17223133146762848, 'eval_accuracy': 0.9235, 'eval_
runtime': 0.8678, 'eval_samples_per_second': 2304.738, 'eval_steps_per_
second': 72.599, 'epoch': 5.0}
```

This gives us an accuracy of 92.35%, which is impressive. A result this strong indicates that the model can reliably identify the predominant emotion present in the text, making it suitable for real-world applications like sentiment analysis.

Model Inference

Finally, we want to be able to use our fine-tuned model for predicting on new, unseen data. The Hugging Face transformers library provides a powerful, high-level API called **pipeline** that makes this incredibly simple.

A pipeline is a high-level wrapper that bundles together a model and its corresponding tokenizer to perform the specified task, abstracting away much of the complexities of inference code. When we create a text classification pipeline with our fine-tuned model, it automatically handles all the necessary steps for us: it takes our raw text string, tokenizes it correctly, feeds the numerical inputs to the model, gets the raw output scores (logits), and then post-processes them into a human-readable format, including the predicted label name and its confidence score.

```
from transformers import pipeline

classifier = pipeline(
    "text-classification",
    model=model,
    tokenizer=tokenizer,
    device=0
)

# 2. Let's test it with some new sentences
custom_text_1 = "I am so excited for this concert tonight, I've been waiting for it for months!"
custom_text_2 = "I'm not sure about this new project. The deadline is so tight and I feel anxious."

# 3. Get predictions
preds = classifier([custom_text_1, custom_text_2])
```

```
for text, result in zip([custom_text_1, custom_text_2], preds):
    print(f"Text: \"{text}\"")
    print(f"  -> Predicted Emotion: {result['label']} (Score:
{result['score']:.4f})")
    print("-" * 50)
```

Text: "I am so excited for this concert tonight, I've been waiting for it for months!"
  -> Predicted Emotion: joy (Score: 0.9945)

Text: "I'm not sure about this new project. The deadline is so tight and I feel anxious."
  -> Predicted Emotion: fear (Score: 0.9976)

The model accurately identifies the emotions in both our sentences with high confidence, demonstrating the success of our fine-tuning process.

Summarization

Our next area of focus is one of the most widely used applications in natural language processing: **automated text summarization**. In the current era of information overload and short attention spans, the ability to automatically condense long texts into short, easy-to-digest summaries is a critical skill for machines to have. The goal is to create a shorter version of a document that captures its most important points, saving time and making complex topics easier to understand.

There are two broad categories of text summarization tasks: extractive and abstractive summarization.

Extractive Summarization

This approach aims to generate a summary by simply extracting some sentences directly from the original text. You can think of it like highlighting sentences while reading a textbook.

Usually, this is achieved by the model scoring each sentence based on its importance and then picking the top-ranking ones to be included in the final summary. Although this method usually produces a factually correct summary, it might sometimes feel disconnected or awkward since it only selects sentences directly from the text.

Abstractive Summarization

This more advanced method aims to summarize a document just like a human would. The model goes through the entire text and then generates new sentences from scratch, summarizing the key information in its own words.

This usually requires the model to truly comprehend the source text so that it can summarize the essence of the text fluently. The summaries produced feel much more natural, just like a human would write. However, note that this is a much harder task than extractive summarization, and it comes with the risk of the model "hallucinating," wherein it might confidently pass off plausible-sounding but incorrect details not found in the given text.

We shall now get our hands dirty with two projects on summarization. In our first project, we shall work on summarizing a Wikipedia article using a pretrained model from the Hugging Face Hub. In our second project, we will work with a dataset containing short conversational chats, where we will go deeper and fine-tune some pretrained models on the Hugging Face Hub.

Summarization Project 1: Summarizing a Wikipedia Article

Our first summarization project applies a pretrained summarization model to generate an abstractive summary of any Wikipedia entry. This example highlights both the ease of using Hugging Face pipelines and the limitations of transformer token windows on long documents. We shall also demonstrate a chunking solution to get around this limitation.

The Wikipedia API

We will first learn how to access Wikipedia articles via the Wikipedia Python library. You may have to install the Wikipedia Python library first:

```
pip install wikipedia
```

You can then access the page corresponding to any topic. For example, here we look at the page on "elephant":

```
import wikipedia

topic = "Elephant"
page = wikipedia.page(topic)
page_content = page.content
print(page_content)
```

Loading a Hugging Face Pretrained Model

Selecting a Model

To select a model, you can go to https://huggingface.co/models, which lists all the pretrained models available on Hugging Face. The left-hand side pane lists all the possible tasks on which you can filter these models. Under "Natural Language Processing", select "Summarization". This will list all the models along with different parameters like number of downloads, number of likes, etc. Let us select the popular facebook/bart-large-cnn model for our use case.

Loading the Model

We can download the facebook/bart-large-cnn model for summarization from the Hugging Face Hub:

```
from transformers import pipeline
summarizer = pipeline("summarization", model="facebook/bart-large-cnn")
```

Using the Model for Summarization

Now, all it takes to get a summary of the Wikipedia article we accessed is to call this summarizer, with the min and max lengths of summary that we are willing to accept:

```
summarizer(page_content[:4000], min_length = 100, max_length = 500)
```

This gave me the following summary text:

```
>>> [{'summary_text': 'Elephants are the largest living land animals. Three
living species are currently recognised: the African bush elephant, the
African forest elephant, and the Asian elephant. Distinctive features of
```

elephants include a long proboscis called a trunk, tusks, large ear flaps, pillar-like legs, and tough but sensitive grey skin. African elephants have larger ears and concave backs, whereas Asian elephants have smaller ears and convex or level backs. They communicate by touch, sight, smell, and sound; elephants use infrasound and seismic communication over long distances.'}]

Note that this model accepts a maximum of 1024 tokens at a time, so we had to select the first 4000 characters of the string "page_content" to make the model work.

Note that the BART model we used has a maximum context window of 1024 tokens, so we had to select the first 4000 characters of the string "page_content" to make the model work. However, slicing the first 4000 characters is a roundabout way to stay under this limit, since characters don't map to tokens directly. A more precise method involves tokenizing the text and then slicing the token IDs, which we will explore in the next subsection, where we discuss an advanced chunking solution.

Chunking to Summarize Large Documents

A significant limitation when working with transformer models is their fixed **context window**. For instance, the facebook/bart-large-cnn model can only process a maximum of 1024 tokens at a time. Since our Wikipedia article is much longer than that, we need a strategy to summarize the entire document.

The most common solution, which we will implement here, is to chop the document into smaller, overlapping chunks and summarize each one individually. We will then stitch these individual summaries together to create a final, comprehensive summary of the entire article.

```
from transformers import AutoTokenizer, pipeline
tokenizer = AutoTokenizer.from_pretrained("facebook/bart-large-cnn")
summarizer = pipeline("summarization", model="facebook/bart-large-cnn")

chunk_size = 1000
overlap_len = 50

token_input_ids = tokenizer(page_content, return_tensors='pt')['input_ids']

chunk_start = 0
total_tokens = len(token_input_ids[0])

print(f'Input text has a total of {total_tokens} tokens.')
```

```
summary_text = ""
while chunk_start < total_tokens:
    chunk_end = min(chunk_start + chunk_size, total_tokens)
    chunk = token_input_ids[:, chunk_start:chunk_end]
    chunk_text = tokenizer.decode(chunk[0], skip_special_tokens=True)
    chunk_summary = summarizer(chunk_text, min_length=10, max_length=500)
    summary_text += chunk_summary[0]["summary_text"] + "\n"
    chunk_start += chunk_size - overlap_len

print(summary_text)
```

Here is the logical flow our code uses to summarize the entire document:

1. **Tokenize the Entire Document:** First, we pass the entire Wikipedia page to the tokenizer. This converts the text into a long sequence of `input_ids`, allowing us to precisely manage the chunks based on token count instead of character count.

2. **Iterate in Overlapping Chunks:** Our `while` loop then slides across this sequence of tokens. For each step, it creates a chunk of 1000 tokens. To avoid losing important context that might occur at the boundary between two chunks, we make every chunk have an overlap of 50 tokens with its previous chunk.

3. **Decode:** Inside the loop, we use `tokenizer.decode()` to convert the numerical token chunk back into a text string. This step is necessary because the `pipeline` is a high-level tool designed to work with raw text; it handles its own internal tokenization.

4. **Summarize:** We call the `summarizer` on this text chunk.

5. **Concatenate the Results:** Finally, the summary from each chunk is appended to our main `summary_text`, building up the full summary of the document piece by piece.

While this method is highly effective, more complex chunking strategies have also been explored. In case of long documents, instead of just outputting the concatenation of all the summaries, we could recursively call the summarizer again on this concatenation and so on until we obtain a summary of the desired length.

Summarization Project 2: Conversational Chat Summarizer

In this project, we will focus again on **abstractive summarization**. Our goal is to create a specialized tool that can summarize short informal chat conversations. For a task like this, the extractive method would fare poorly; simply pulling out sentences like "Hey!" or "lol, sure" would create a nonsensical summary. We need a model that can understand the *intent* of the conversation and report on its conclusions, such as "Amanda and Jerry agreed to meet at the bookstore tomorrow."

Google's T5 Model

To achieve this level of comprehension, we will use Google's T5 model, which is an encoder-decoder transformer specialized for sequence-to-sequence tasks. This model treats every NLP problem as text generation, requiring short prefixes to indicate the task to be performed (e.g., "summarize" or "translate English to French"). This makes it especially suitable for text generation tasks like summarization, translation, question answering, etc.

We fine-tune T5 on the **SAMSum dataset** (curated by Samsung Research[1]), a collection of thousands of real-world dialogues. Through this process, we will transform a generalist language model into a highly effective specialist capable of making sense of informal conversations.

Imports

You may need to install the following packages first in your notebook, depending on your existing versions of these different packages:

```
!pip install datasets==3.6.0
!pip install evaluate rouge_score
```

I have demonstrated the use of pip here, but you can use any other package manager that suits your system.

---

[1] https://arxiv.org/abs/1911.12237v2

You will need the following imports.

```
import datasets
import huggingface_hub
import fsspec

import evaluate
import numpy as np
from transformers import (
    AutoModelForSeq2SeqLM,
    DataCollatorForSeq2Seq,
    Seq2SeqTrainingArguments,
    Seq2SeqTrainer
)
from transformers import pipeline
```

Dataset

We use Hugging Face's load_dataset function to load the SAMSun dataset.

```
from datasets import load_dataset

raw_dataset = load_dataset("knkarthick/samsum")
print(raw_dataset)
```

Exploratory Data Analysis (EDA)

Let us first look at an arbitrary example in the dataset:

```
sample = raw_dataset["train"][9]

print("--- DIALOGUE ---")
print(sample["dialogue"])
print("\n--- SUMMARY ---")
print(sample["summary"])
```

This prints out the following.

```
--- DIALOGUE ---
Matt: Do you want to go for date?
Agnes: Wow! You caught me out with this question Matt.
```

Matt: Why?
Agnes: I simply didn't expect this from you.
Matt: Well, expect the unexpected.
Agnes: Can I think about it?
Matt: What is there to think about?
Agnes: Well, I don't really know you.
Matt: This is the perfect time to get to know eachother
Agnes: Well that's true.
Matt: So let's go to the Georgian restaurant in Kazimierz.
Agnes: Now your convincing me.
Matt: Cool, saturday at 6pm?
Agnes: That's fine.
Matt: I can pick you up on the way to the restaurant.
Agnes: That's really kind of you.
Matt: No problem.
Agnes: See you on saturday.
Matt: Yes, looking forward to it.
Agnes: Me too.

--- SUMMARY ---
Matt invites Agnes for a date to get to know each other better. They'll go to the Georgian restaurant in Kazimierz on Saturday at 6 pm, and he'll pick her up on the way to the place.

Let us now print out some statistics related to the dataset.

```
# Convert the dataset splits to pandas DataFrames
train_df = pd.DataFrame(raw_dataset["train"])
test_df = pd.DataFrame(raw_dataset["test"])

# Calculate the number of words in each dialogue and summary
train_df['dialogue_length'] = train_df['dialogue'].str.split().str.len()
train_df['summary_length'] = train_df['summary'].str.split().str.len()
```

```python
# Get descriptive statistics
print("Statistics for Text Lengths in the Training Set:")
print(train_df[['dialogue_length', 'summary_length']].describe())

# Plot the distributions
fig, axes = plt.subplots(1, 2, figsize=(12, 5))
sns.histplot(train_df['dialogue_length'], bins=50, ax=axes[0],
color='seagreen')
axes[0].set_title('Distribution of Dialogue Lengths')
axes[0].set_xlabel('Length (Number of Words)')

sns.histplot(train_df['summary_length'], bins=30, ax=axes[1],
color='salmon')
axes[1].set_title('Distribution of Summary Lengths')
axes[1].set_xlabel('Length (Number of Words)')

plt.tight_layout()
plt.show()
```

This prints out the following statistics, along with the histograms in Figure 5-3.

```
Statistics for Text Lengths in the Training Set:
       dialogue_length  summary_length
count     14731.000000    14732.000000
mean         93.792750       20.317472
std          74.031937       11.153815
min           7.000000        1.000000
25%          39.000000       12.000000
50%          73.000000       18.000000
75%         128.000000       27.000000
max         803.000000       64.000000
```

CHAPTER 5   PRACTICAL NATURAL LANGUAGE PROCESSING WITH HUGGING FACE

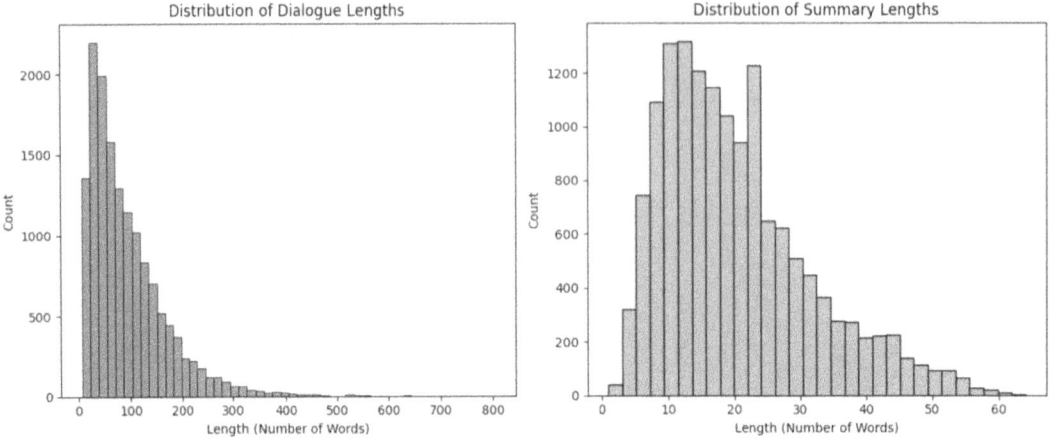

Figure 5-3. Training data dialogue and summary lengths in the training data

Let's analyze again this line: train_df['dialogue_length'] = train_df['dialogue'].str.split().str.len().

First, the .str.split() method is applied to each dialogue string, converting it to a list of individual words. Then, the .str.len() method is applied to this new series of lists, calculating the length of each list to get the final word count.

Looking at the statistics, you may have noticed that the summary_length has one sample more than the dialogue_length. Let us try to find out the null dialogue that may be causing this.

```
train_df = pd.DataFrame(raw_dataset["train"])
problematic_rows = train_df[train_df['dialogue'].isnull()]

print("Found the row with the missing dialogue:")
print(problematic_rows)
Found the row with the missing dialogue:
            id dialogue                                    summary
6054  13828807     None  problem with visualization of the content
```

This detects the culprit "None" dialogue.

Tokenization and Preprocessing

After our EDA, we preprocess the data into a numerical format that a model like T5 can understand. Toward that, we now define our preprocess function.

```python
from transformers import AutoTokenizer

# 1. Load the tokenizer for our chosen model
model_checkpoint = "t5-small"
tokenizer = AutoTokenizer.from_pretrained(model_checkpoint)

def preprocess_function(examples):
    # Add the T5 prefix for the summarization task
    prefix = "summarize: "
    inputs = [prefix + doc for doc in examples["dialogue"]]

    # Tokenize the input dialogues
    model_inputs = tokenizer(inputs, max_length=1024, truncation=True)

    # Tokenize the summaries (the labels/targets)
    with tokenizer.as_target_tokenizer():
        labels = tokenizer(examples["summary"], max_length=128,
            truncation=True)

    # Add the tokenized labels to our model inputs
    model_inputs["labels"] = labels["input_ids"]
    return model_inputs

print("Original dataset length:", len(raw_dataset["train"]))

def is_not_empty(example):
    return example["dialogue"] is not None and example["summary"] is
    not None

cleaned_dataset = raw_dataset.filter(is_not_empty)

print("Cleaned dataset length:", len(cleaned_dataset["train"]))

tokenized_datasets = cleaned_dataset.map(preprocess_function, batched=True)

print("\n--- After successful tokenization ---")
print(tokenized_datasets)
```

CHAPTER 5   PRACTICAL NATURAL LANGUAGE PROCESSING WITH HUGGING FACE

The above code block tokenizes our text while demonstrating several key concepts for working with Hugging Face:

1. We first add the prefix "summarize: " to all the input dialogues. This is dictated by a convention followed by the T5 model – it primes the versatile T5 model for the summarization task.

2. The "`with tokenizer.as_target_tokenizer()`" is a vital step for encoder-decoder models, as it ensures the `labels` are formatted correctly for the decoder component, which is essential for training to work properly.

3. Before tokenization, we apply a `filter()` function to remove the faulty rows with empty dialogues (or summaries) we discovered during our EDA.

4. Finally, we apply our preprocessing function using Hugging Face's "map" function. Processing in batches using map is significantly faster and more scalable than using a simple `for` loop.

Evaluation Metric

The ROUGE Score

The standard metric for evaluating a model-generated summary against a reference summary is **ROUGE**, which stands for **Recall-Oriented Understudy for Gisting Evaluation**. ROUGE scores a (model-generated) summary based on the overlap of words and word sequences between it and the reference summary, effectively measuring how well the model captured the key information present in the reference.

While this approach of simply counting overlapping words may seem like a crude measure (as it cannot truly gauge semantic meaning or grammatical correctness), ROUGE has proven to be a surprisingly effective and reliable proxy for human evaluation, making it the industry standard.

The ROUGE metric is not a single score but a suite of related metrics, each measuring a different aspect of this overlap. The most commonly used ones are **ROUGE-1**, which measures the overlap of individual words (unigrams), and **ROUGE-2**, which measures the overlap of word pairs (bigrams). While ROUGE-1 assesses content coverage, ROUGE-2 is a stricter measure of fluency and phraseology. Another important variant is **ROUGE-L**, which finds the longest common subsequence of words between the generated and reference summaries. This score rewards models that not only use the right words but also maintain a similar sentence structure and word order to the human-written summary.

We will now load the rouge_metric from the evaluate module for our project.

```
rouge_metric = evaluate.load("rouge")
```

The Compute_metrics Function

We will now define the compute metrics function, which we will provide to the Hugging Face Trainer for computing our metrics of interest.

```
def compute_metrics(eval_pred):
    predictions, labels = eval_pred

    # Clean both predictions and labels by replacing -100 with the pad token ID
    predictions = np.where(predictions != -100, predictions, tokenizer.pad_token_id)
    labels = np.where(labels != -100, labels, tokenizer.pad_token_id)

    decoded_preds = tokenizer.batch_decode(predictions, skip_special_tokens=True)
    decoded_labels = tokenizer.batch_decode(labels, skip_special_tokens=True)

    # ROUGE expects a newline after each sentence
    decoded_preds = ["\n".join(pred.strip().split()) for pred in decoded_preds]
    decoded_labels = ["\n".join(label.strip().split()) for label in decoded_labels]

    # Compute ROUGE scores
    result = rouge_metric.compute(predictions=decoded_preds, references=decoded_labels, use_stemmer=True)

    # Extract the median scores
    result = {key: value * 100 for key, value in result.items()}
```

```
# Add a measure of prediction length
prediction_lens = [np.count_nonzero(pred != tokenizer.pad_token_id) for
pred in predictions]
result["gen_len"] = np.mean(prediction_lens)

return {k: round(v, 4) for k, v in result.items()}
```

This function is a crucial component of our training and evaluation processes and deserves a step-by-step analysis.

The function receives a single argument, eval_pred, which is a tuple containing the model's raw token ID predictions and the true label token IDs. We start by unpacking this tuple into two separate arrays.

```
predictions = np.where(predictions != -100, predictions, tokenizer.
pad_token_id)
labels = np.where(labels != -100, labels, tokenizer.pad_token_id)
```

The Trainer uses the special value -100 to mark tokens that should be ignored during loss calculation (like padding). However, our tokenizer doesn't know what -100 means and will error out if it tries to decode it. This code uses NumPy's where function to find every -100 in both arrays and replace it with the tokenizer's actual padding token ID (tokenizer.pad_token_id). This makes the arrays "safe" for decoding.

```
decoded_preds = tokenizer.batch_decode(predictions, skip_special_
tokens=True)
decoded_labels = tokenizer.batch_decode(labels, skip_special_
tokens=True)
```

Now that our arrays are clean, we use tokenizer.batch_decode() to convert the numerical token IDs back into human-readable text strings. The skip_special_tokens=True argument conveniently removes any remaining special tokens (like the padding token we just added) from the final text.

```
decoded_preds = ["\n".join(pred.strip().split()) for pred in
decoded_preds]
decoded_labels = ["\n".join(label.strip().split()) for label in
decoded_labels]
```

The standard implementation of the ROUGE metric requires each sentence in the text to be separated by a newline character. This list comprehension first strips any leading/trailing whitespace (`.strip()`), splits the summary into a list of words (`.split()`), and then joins them back together with a newline (`"\n".join(...)`). This ensures our summaries are in the exact format the metric expects.

```
result = rouge_metric.compute(predictions=decoded_preds,
references=decoded_labels, use_stemmer=True)
```

Here, we pass our cleaned and formatted predictions and labels to the rouge_metric object we loaded earlier. It compares the two sets of summaries and calculates the ROUGE-1, ROUGE-2, and ROUGE-L scores, returning them in a dictionary.

```
result = {key: value * 100 for key, value in result.items()}

prediction_lens = [np.count_nonzero(pred != tokenizer.pad_token_id) for
pred in predictions]
result["gen_len"] = np.mean(prediction_lens)
```

For easier readability, we multiply the raw ROUGE scores (which are between 0 and 1) by 100 to express them as percentages. We also add a new, useful metric to our results: gen_len, which contains the average length of the summaries our model generated. This is helpful for diagnosing issues arising due to the summaries being too long or too short.

Finally, we return this dictionary after rounding all its values to four decimal places.

Model Training

We will now load the pretrained T5-small model and fine-tune it for our data.

```
# --- Load the Model ---
model_checkpoint = "t5-small"
model = AutoModelForSeq2SeqLM.from_pretrained(model_checkpoint)

# --- Define the Data Collator ---
data_collator = DataCollatorForSeq2Seq(tokenizer=tokenizer, model=model)

training_args = Seq2SeqTrainingArguments(
    output_dir="samsum_t5_finetuned",
    learning_rate=2e-5,
    per_device_train_batch_size=16,
```

```
    per_device_eval_batch_size=16,
    weight_decay=0.01,
    save_total_limit=3,
    num_train_epochs=3,
    predict_with_generate=True,
    fp16=True,
    report_to="none",
    generation_max_length=50,
    generation_num_beams=4,
)

# --- Create the Trainer ---
trainer = Seq2SeqTrainer(
    model=model,
    args=training_args,
    train_dataset=tokenized_datasets["train"],
    eval_dataset=tokenized_datasets["validation"],
    tokenizer=tokenizer,
    data_collator=data_collator,
    compute_metrics=compute_metrics,
)

# --- Start Training ---
print("Starting the fine-tuning process...")
trainer.train()
```

The above code concisely sets up our entire fine-tuning process. Let's understand the key components that make this process so easy to set up:

1. **AutoModelForSeq2SeqLM**: We load our pretrained T5 model using this class. As the name suggests, it's an "auto" class designed to load any compatible **sequence-to-sequence** model architecture. This is a suitable choice for tasks like summarization or translation, where the model reads an input sequence and has to generate a new output sequence.

2. **DataCollatorForSeq2Seq**: As we saw before, a data collator organizes the input data into batches before feeding it to the model. For this, we use the specialized DataCollatorForSeq2Seq because it correctly pads our input dialogues and output summaries, which can have different lengths. Most importantly, it also prepares the labels correctly for the decoder part of our T5 model, a critical step for the training to work.

3. **Seq2SeqTrainingArguments**: The Seq2SeqTrainingArguments object is where we define all our training parameters, from the learning rate to evaluation settings.

4. **Seq2SeqTrainer**: We pass everything – the model, arguments, data collator, datasets, and the metrics calculator function – to the Seq2SeqTrainer. The **Trainer** then automates the entire training loop for us, including moving data to the GPU, running evaluations, and saving our final model, which saves us from writing a great deal of boilerplate code.

We used a lot of training arguments in our code. Let us see what each of them means.

1. **output_dir**: Specifies the directory where the fine-tuned model checkpoints and final model files will be saved.

2. **learning_rate**: Sets the initial learning rate for the AdamW optimizer. Note that we intentionally used a very small learning rate (2e-5) for fine-tuning, since a higher learning rate can destroy the model's valuable pretrained knowledge and lead to poor performance.

3. **per_device_train_batch_size and per_device_eval_batch_size**: Define the number of training examples and evaluation examples, respectively, to be processed in a single batch on one GPU.

4. **weight_decay**: Applies a small regularization penalty to the model's weights to help prevent overfitting.

5. **save_total_limit**: Limits the number of checkpoints saved. With our setting, it will only keep the most recent three.

6. **num_train_epochs**: Sets the total number of epochs over which the model will train.

7. **predict_with_generate**: This crucial argument instructs the Trainer to generate actual text summaries during evaluation, which is essential for calculating metrics like ROUGE.

8. **fp16**: Enables mixed-precision training, which uses a combination of 16-bit and 32-bit floating points to significantly speed up training and reduce memory usage on compatible GPUs.

9. **report_to**: Controls where training logs are sent; we set it to "none" to disable integrations with any external services (like wandb).

10. **generation_max_length**: Sets the maximum token length for summaries generated during evaluation, ensuring a consistent and fair test.

11. **generation_num_beams**: Activates beam search during text generation, a more advanced decoding strategy that often results in higher-quality output.

Model Testing and Inference

Testing the model is as simple as calling the evaluate function on the Trainer:

```
test_results = trainer.evaluate(eval_dataset=tokenized_datasets["test"])
print(test_results)
```

This single line automatically orchestrates the entire evaluation loop for us. Behind the scenes, the Trainer handles a series of critical steps: it moves our test data to the GPU and iterates batchwise through every example. For each dialogue, it generates a summary using the generation settings we defined in our training arguments. It then passes the generated summaries and the reference summaries to our custom compute_metrics function to calculate the final ROUGE scores. The test_results object that is returned is a dictionary containing the final, aggregated metrics for our unseen test data.

Results

We get the following metrics for our test set:

```
{'eval_loss': 1.8364745378494263,
 'eval_model_preparation_time': 0.003,
 'eval_rouge1': 43.6601,
 'eval_rouge2': 19.3357,
 'eval_rougeL': 35.4198,
 'eval_rougeLsum': 43.6752,
 'eval_gen_len': 47.105,
 'eval_runtime': 52.6836,
 'eval_samples_per_second': 15.546,
 'eval_steps_per_second': 0.987}
```

The most important scores here are for the ROUGE metric: a **ROUGE-1 of 43.66** and a **ROUGE-2 of 19.33**. These are solid results[2] for a t5-small model after a brief fine-tuning process, indicating that our model is effectively capturing both the key terms and the phraseology of the reference summaries.

Inference

We will now use our model to create a pipeline object, which we will use for inferencing on an unseen dialogue.

```
summarizer = pipeline(
    "summarization",
    model=model,
    tokenizer=tokenizer,
    device=0,  # Use 0 for GPU, -1 for CPU
    min_length=15,
    max_length=50,
)
# Prepare a new, unseen dialogue to summarize
unseen_dialogue = """
```

---

[2] For reference ROUGE scores, see the Hugging Face model cards for T5-small and bart-large fine-tuned on the SAMSum dataset.

```
Leo: Hey Mia, are you still planning on hitting the gym after work?
Mia: Definitely! I was thinking around 6 PM. Does that work for you?
Leo: 6 PM is perfect. I'll meet you by the weight racks.
Mia: Sounds good, see you there!
"""

# Use the pipeline to generate the summary
result = summarizer(unseen_dialogue, min_length=15,
    max_length=50, num_beams=4)
                    # , no_repeat_ngram_size=2)

# Print the result
print("\n--- UNSEEN DIALOGUE ---")
print(unseen_dialogue)
print("\n--- MODEL-GENERATED SUMMARY ---")
print(result[0]['summary_text'])
```

This generates the summary:

```
--- MODEL-GENERATED SUMMARY ---
Leo and Mia are planning on hitting the gym after work around 6 PM. Mia
will meet Mia by the weight racks.
```

Note that the summary is on the right track, but not perfect – "Mia will meet Mia" spoils it. You can try uncommenting the line "no_repeat_ngram_size=2)" in the summarizer call (and remove the extra parenthesis), and that improves the summary slightly:

```
Leo is planning on hitting the gym after work around 6 PM. Mia will meet
her by the weight racks. She's going to meet him at 6 pm.
```

This is still far from perfect, but better than before. Setting the "no_repeat_ngram_size" to 2 tells the model to avoid repeating any 2-gram in the summary. You can try playing around with the other parameters here to help improve the model performance even further: https://huggingface.co/docs/transformers/main_classes/text_generation#transformers.GenerationConfig.

Question Answering

We will now work on another useful NLP task: **Extractive Question Answering (QA)**. The goal here is to build a system that can find the precise answer to a specific question from within a given text document. This is far more advanced than a simple keyword search; the model must understand the semantics of both the question and the provided context to pinpoint the exact span of text that contains the answer.

We will use a powerful pretrained model from the Hugging Face Hub to quickly build a "smart search" tool that can answer simple questions about a given paragraph.

We will first load the question answering pipeline.

```
from transformers import pipeline

# 1. Load the QA pipeline
model_name = "deepset/roberta-base-squad2"
qa_pipeline = pipeline(
    "question-answering",
    model=model_name,
    tokenizer=model_name
)
```

The question answering model needs a context from which it picks its answers. Furthermore, the model RoBERTa, which we have loaded here, has a limited context window of 512 tokens, which means that the context, along with the question, needs to be limited to a maximum of 512 tokens. So we will define a short context and then ask the model to find answers to our three questions from it. We will then pass these questions along with the context to the pipeline and print out the answers along with the accompanying confidence scores.

```
# 2. Define a short, specific context that can be processed in one go.
context = """
Jupiter is the fifth planet from the Sun and the largest in the Solar System.
It is a gas giant with a mass more than two and a half times that of all
the other planets in the Solar System combined. Mars is the fourth planet
from the Sun and the second-smallest planet, often referred to as the 'Red
Planet' due to its reddish appearance. Saturn, the sixth planet, is another
gas giant best known for its stunning ring system.
"""
```

```
# 3. Create a list of questions to ask about the context.
questions = [
    "Which planet is known as the Red Planet?",
    "Which is the largest planet in the solar system?",
    "Which planet is best known for its ring system?"
]

print("\n--- Asking Questions ---")

# 4. Get an answer for each question
for question in questions:
    result = qa_pipeline(question=question, context=context)

    # 5. Print the formatted result for each question.
    print(f"\nQ: {question}")
    print(f" A: {result}")
```

Q: Which planet is known as the Red Planet?
 A: {'score': 0.7733553051948547, 'start': 200, 'end': 204, 'answer': 'Mars'}

Q: Which is the largest planet in the solar system?
 A: {'score': 0.7992478013038635, 'start': 1, 'end': 8, 'answer': 'Jupiter'}

Q: Which planet is best known for its ring system?
 A: {'score': 0.7099953889846802, 'start': 340, 'end': 346, 'answer': 'Saturn'}

Our pipeline gives us all the correct answers, along with a score between 0 and 1, indicating the confidence our model has about each answer. It also pinpoints the location in the context where that answer can be found.

Summary

- The Hugging Face Hub provides access to thousands of pretrained models that serve as effective starting points for almost any NLP task, allowing for rapid prototyping and establishing baselines. Alongside these models, the Hub also hosts a vast collection of datasets that can be loaded and used with just a few lines of code.

- The `pipeline` API provides a simple way to use a **pretrained** model for a specific task, abstracting away the complexities of tokenization and post-processing.

- One of the commonly used techniques in modern NLP is **fine-tuning**, which involves adapting a generalized pretrained model to become a specialist for the task at hand.

- **Pretrained models** are ideal for getting started quickly, whereas **fine-tuning** is the key to unlocking state-of-the-art performance for a specific use case.

- The `Trainer` class is a high-level API that automates the entire fine-tuning loop by handling device placement, batching, backpropagation, and evaluation, saving us from writing extensive boilerplate code.

- The standard workflow for data preprocessing using Hugging Face libraries involves using the `datasets` library to load and analyze data and an `AutoTokenizer` with the `.map()` method to efficiently convert text into a model-ready tokenized (numerical) format.

- Since transformer models have a fixed context window, processing long documents gets challenging; one strategy we used in our Wikipedia article summarization project was to break the text into smaller, overlapping chunks that could be processed individually, and the results were concatenated together to get the final summary.

- The quality of generated text can be significantly improved by moving beyond default settings and using decoding strategies like beam search and other parameters to control repetition, as we saw in our second summarization project.

CHAPTER 5   PRACTICAL NATURAL LANGUAGE PROCESSING WITH HUGGING FACE

- Different NLP tasks require specific components; for example, sequence-to-sequence tasks like summarization need a DataCollatorForSeq2Seq and special prefixes, while classification requires a DataCollatorWithPadding.

- You can classify text into custom categories without any training by using a zero-shot-classification pipeline, which cleverly repurposes a Natural Language Inference (NLI) model for on-the-fly classification.

- The question-answering pipeline we saw in our last project enables Extractive Question Answering, a powerful technique where the model comprehends the semantics of the question and the provided context and extracts the exact span of text containing the answer from within the context.

CHAPTER 6

Building a Language Model for Storytelling

In this chapter, we work on a project aimed at building a storytelling language model from scratch in PyTorch. We start by explaining the foundations of language models, describing some architectures historically used in this regard. We then deep dive into the mathematics of the transformer architecture, which we will use later in our project.

We then start with our language modeling project, where we begin with character-level tokenization. We first implement a simple **bigram model** to serve as a performance baseline. Subsequently, we build a **decoder-only transformer**, writing modular code for each component like the self-attention mechanism. Finally, we switch to **word-level tokenization** to demonstrate a significant improvement in the coherence of the generated results.

Introduction to Language Modeling

Language modeling is the task of predicting the next word in a sequence. Language models assign probabilities to every possible next word in a sequence; consequently, they also assign probabilities to entire sentences. This is a foundational task for NLP, since a model that can accurately predict the next word in a sentence must, by necessity, have learned a great deal about the grammar, syntax, semantics, and even the common-sense "world knowledge" embedded in the language. This predictive capability is the engine that powers a vast array of NLP applications, from machine translation and text summarization to the conversational abilities of AI chatbots.

More formally, given a text sequence <$w_1, w_2, ..., w_{m-1}$>, a language model aims to predict the probability of the next word: $\Pr[w_m|w_1, ..., w_{m-1}]$ for all possible words w_m in the vocabulary. The model is trained to maximize the likelihood of the entire training corpus, meaning it learns its parameters so as to maximize the probabilities of the sequences in its training data.

The approaches to language modeling have evolved significantly over time, with each new paradigm seeking to overcome the limitations of its predecessor.

N-Gram Models

Early approaches to language modeling were purely statistical. One such popular approach was that of the n-gram model. It was based on the idea of approximating the probability of a word occurring next based on only the previous $n - 1$ words instead of the entire preceding sequence.

For example, when n is 2, the resulting model is termed the bigram model. The bigram model estimates the probability of the next word by simply the probability of its occurrence given the previous word. For example, in the sentence "I plucked the red …", to estimate the probability of "rose" being the next word, the bigram model would simply estimate the probability of the word "rose" succeeding the word "red", ignoring all the previous context. More formally, it approximates

$$\Pr[w_m|w_1,...,w_{m-1}] \approx \Pr[w_m|w_{m-1}]$$

These models, as you may have guessed, are too simple to generalize. They are dependent on a limited context, rendering them incapable of learning longer-range dependencies, like a subject and its verb separated by a long clause.

Deep Learning Models: RNNs and LSTMs

The first major shift in overcoming these issues came with **recurrent neural networks (RNNs)**. Unlike n-grams, which have a fixed context window, an RNN (as we saw in Chapter 4) is designed to handle arbitrary-length sequences. To recap, it processes tokens sequentially, capturing a hidden state at each step, which it passes along to the next step along with the next input token. On a high level, this allows the model to memorize some of its context, which can be used when processing later parts of the sequence.

It processes text one token at a time, and after each step, it passes along a **hidden state** – a vector representation that acts as a form of memory – to the next step. In theory, this allowed the model to maintain context from the very beginning of a sequence when making a prediction far down the line. An improvement over RNNs is the long short-term memory (LSTM) architecture, which contains additional gates controlling which information can be read, written, and erased from its memory.

The Magic of Transformers

The major downside of recurrent neural network models is their inherent sequentiality. The hidden state for time step t cannot be computed until the processing for time step t-1 is complete. This bottleneck was shattered by the introduction of the transformer architecture. The transformer completely dispenses with recurrence and instead relies on an intelligent mechanism called **attention**, which allows it to process all tokens in a sequence simultaneously. This ability to capture complex, long-range dependencies while being highly parallelizable has made the transformer the undisputed state-of-the-art for virtually all modern language modeling tasks, forming the basis of cutting-edge systems like ChatGPT.

The main focus of this chapter will be building a language model using the transformer decoder-only architecture. As such, we will now proceed to study this decoder-only transformer architecture in greater detail.

Transformer Decoder-Only Architecture

We will begin by studying the self-attention mechanism, which is at the heart of the transformer architecture.

Self-attention Mechanism

As we learnt briefly in Chapter 4, self-attention enables every token in a sequence to interact with every other previous token and weigh its relevance. To achieve this, each input token's vector representation is projected into three distinct, learnable roles:

- **Query (Q):** A vector representing what a token is "looking for" or "asking about." This is used for the token currently under consideration.

- **Key (K):** A vector representing what information a token "offers" or what it can be identified by. This is used for every token preceding the token under consideration (and including itself), to compare it with the current token's query representation.

- **Value (V):** A vector representing the actual content or substance of the token. This is the vector contributed to by every preceding element in the weighted sum to obtain the output for the current token.

Let us now formalize this concept mathematically. Consider an input vector of integer tokens, tokenized using a vocabulary as $x = (x_1, x_2, ..., x_m)$. We would first use an embedding layer to get an embedding e_i, of dimension d, for each token. We then use linear projections, which are essentially multiplications with projection matrices, to get the corresponding query (q_i), key (k_i), and value (v_i) vectors for each token as

Query Projection: $q_i = Q \cdot e_i = (q_{i1}, q_{i2}, ..., q_{ih})$ for $i \in \{1, ..., m\}$
Key Projection: $k_i = K \cdot e_i = (k_{i1}, k_{i2}, ...k_{ih})$ for $i \in \{1, ..., m\}$
Value Projection: $v_i = V \cdot e_i = (v_{i1}, v_{i2}, ..., v_{ih})$ for $i \in \{1, ..., m\}$

Here, Q, K, V are matrices of dimension $(d \times h)$ each. They correspond to linear layers with learnable parameters. Here, d is the token embedding dimension, and h is the head size.

Now, let us consider the example where we are looking for tokens relevant to token x_4. Then, we will fetch its query vector q_4 and key vectors for the other tokens before it, **including itself**, i.e., (k_1, k_2, k_3, k_4). We will then compute the raw attention scores by taking the dot product of q_4 with each of these keys. For example, the attention score for token x_4 with respect to token x_1 will be computed as

$$\text{attention-score}(x_4, x_1) = q_4 \cdot k_1$$

Scaled Dot-Product Attention

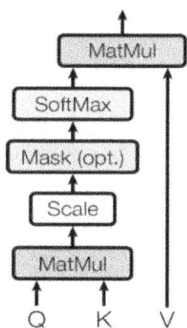

Figure 6-1. *Scaled dot-product attention. Picture credit: "Attention Is All You Need" by Vaswani et al*

Scaled Dot-Product Attention

The magnitudes of these scores would depend heavily on the embedding dimension; to remove this dependence, we scale these raw attention scores by the square root of h, to get the scaled dot-product attention (SDPA) scores as

$$\text{SDPA}(x_4, x_1) = \frac{q_4 \cdot k_1}{\sqrt{\{h\}}}$$

These scaled attention scores give us a measure of how relevant a specific token (x_1, in our case) is to our token under consideration (x_4). We now convert these attention scores into a clean probability distribution by applying a softmax function to them.

$$\text{Attention-weight}(x_4, x_1) = \frac{e^{SDPA(x_4, x_1)}}{\sum e^{SDPA(x_4, x_i)}} = w_{41}$$

We then get the output vector for token x_4 as a weighted average of the values of all the previous tokens: $v_1 w_{41} + v_2 w_{42} + v_3 w_{43} + v_4 w_{44}$. This entire process is depicted in Figure 6-1.

One intuition to understand this application of softmax is as follows: a crude method of using these attention scores would have been to take the token, say x_2, with the **max** attention score for a token (x_4) under consideration, and directly pass its value v_2 as the token x_4's output for the next level. Instead of using such a harsh constraint as max, we instead use the **softmax**, which is a smoother version of max, to weigh the values of tokens according to their relevance for our token x_4.

This process therefore results in a new, context-rich representation for every token, where each token's vector contains a blend of information from all the other tokens preceding it that it paid attention to. The representation of each word gets refined in this way in each self-attention layer.

Multi-head Attention

Instead of relying on a single attention calculation, the transformer employs **multi-head attention** (Figure 6-2). This is like having a panel of experts, where each "head" is a separate query-key-value attention mechanism. This allows the model to learn different aspects of the data in parallel in its different heads – for example, one head might focus on grammatical structure while another tracks semantic meaning. The outputs of all the heads are concatenated and passed through a final linear layer to create a unified representation.

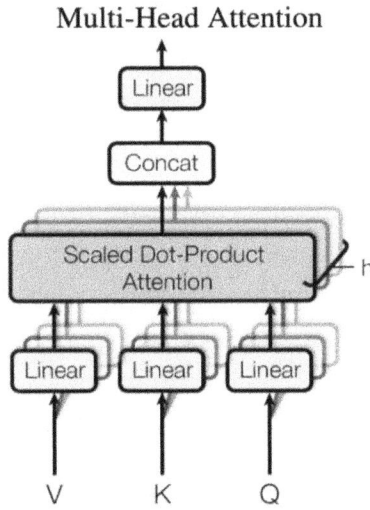

Figure 6-2. *Picture credit: "Attention Is All You Need" by Vaswani et al*

Positional Encodings

The self-attention mechanism is agnostic of the word order. It treats an input sequence like a "bag of words." For example, "the tiger chased the deer" and "the deer chased the tiger" would look nearly identical to the model, even though they have completely opposite meanings.

To address this shortcoming, we create a dedicated vector for each position in the sequence called a **positional embedding**. This positional embedding vector is simply added to the token's own embedding, so that the input now contains information about both the token's identity and its order.

Transformers typically use either fixed positional encodings (e.g., using sine and cosine functions of different frequencies) or learned positional embeddings (vectors learned during the training process). For our project, we use learned positional embeddings for simplicity and effectiveness.

The Decoder Block

All these components are assembled into a **decoder block**, the fundamental, repeatable unit of our model. Each block consists of two main sub-layers: a **multi-head attention** layer and a **feed-forward neural network** (a simple neural network that processes each token individually).

Furthermore, each of these sublayers is wrapped with

1. A **residual connection**, which adds the input of the sublayer to its output (`x + Sublayer(x)`). This acts as an "information short circuit," thereby helping in the training of very deep models.

2. A **layer normalization** step, which stabilizes the training process. Unlike batch normalization, which relies on batch-wide statistics, layer normalization operates independently on each token by computing the mean and standard deviation across its embedding dimensions. This per-token normalization ensures that the input to each sublayer maintains a stable distribution with zero mean and unit variance. This helps in mitigating issues like vanishing or exploding gradients and promotes faster convergence during training.

The Final Transformer Language Model

Having detailed each individual component, we now construct a complete picture of our language model (Figure 6-3). The architecture begins by converting the input tokens into vector representations by summing their **token embeddings** (representing meaning) and **positional embeddings** (representing order). This sequence of vectors is then

CHAPTER 6   BUILDING A LANGUAGE MODEL FOR STORYTELLING

passed through a deep stack of our decoder blocks comprising the multi-head attention layers and feed-forward networks, along with residual connections and layer norms as described above.

The token representations get refined with each pass, as they repeatedly exchange and process information via the self-attention and feed-forward sublayers. After exiting the final block, the resulting vectors are passed through one last **layer normalization** step and then a final **linear layer**, often called the language model head. This layer projects the high-dimensional internal representation back into the vast vocabulary space to produce the final prediction scores, or **logits**, for every possible next token.

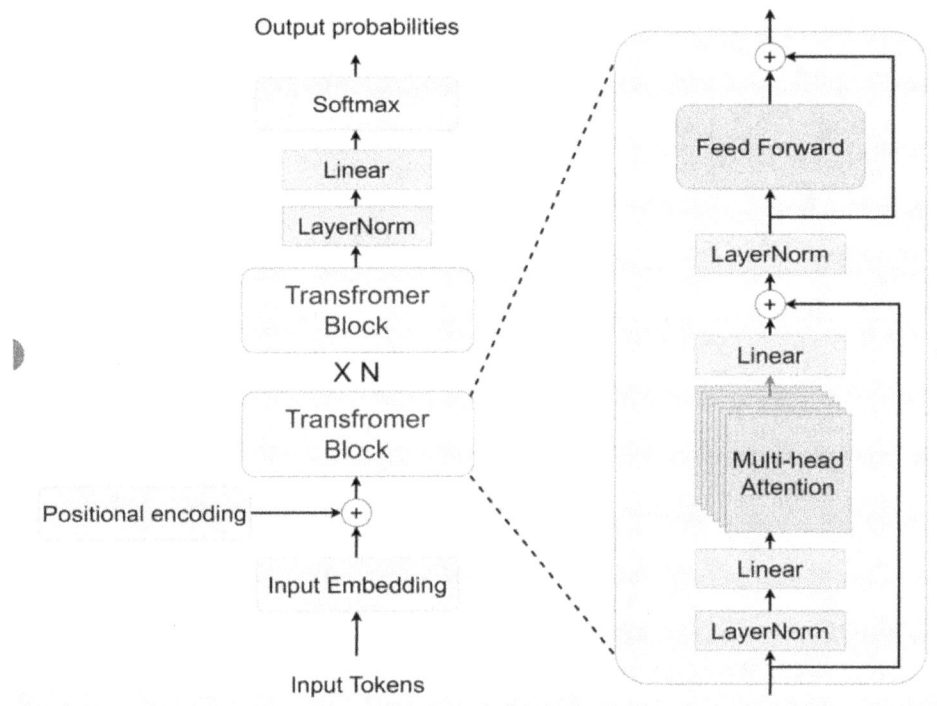

Figure 6-3. Picture credit: https://www.researchgate.net/figure/Decoder-only-Transformer-architecture-The-input-to-the-decoder-is-tokenized-text-and_fig3_373183262

Project: A Story Writing Language Model from Scratch

In this project, we build a language model to generate very short stories for children, training using the TinyStories dataset. An excellent resource in this regard is the "Let's build GPT" tutorial by Andrej Karpathy, which is also one of the inspirations for this project of ours.

A Note on Computing Resources

We recommend using GPUs with large RAMs for these projects, to ensure training completes in a reasonable time. I used an NVIDIA A100 GPU in Google Colab (paid tier) for fast development and experimentation. However, the code is designed to work perfectly on more commonly available GPUs like T4 or L4 or on most modern consumer GPUs. If you don't have access to a GPU, you can still run the code on a CPU, reducing the model size (n_layer, n_embd, etc.), and running it for fewer training iterations as required.

Imports

You may need to install the following version of datasets, depending on your environment.

```
!pip install datasets==3.6.0
```

You will need the following packages imported for this project:

```
import torch
import torch.nn as nn
from torch.nn import functional as F

import time
import math
import random

from dataclasses import dataclass
from datasets import load_dataset
```

CHAPTER 6   BUILDING A LANGUAGE MODEL FOR STORYTELLING

Dataset

To get a simple, yet expressive, language model, we use the **TinyStories** dataset – a charming collection of extremely short, syntactically clean stories written for young children. TinyStories is a dataset specifically designed for training a small language model. Unlike traditional datasets, which are curated by scraping the web, TinyStories is a synthetic corpus, meaning it was generated entirely by a large, advanced AI – in this case, GPT-4.

Each story is just a few sentences long, making the dataset ideal for fast experimentation and quick model iterations. TinyStories is hosted on Hugging Face's datasets library, so loading it is as simple as a single function call.

```
# Loading and Inspecting TinyStories

raw_datasets = load_dataset("roneneldan/TinyStories")

print("--- Dataset Information ---")
print(raw_datasets)
```

This prints out the following information about the dataset.

```
DatasetDict({
    train: Dataset({
        features: ['text'],
        num_rows: 2119719
    })
    validation: Dataset({
        features: ['text'],
        num_rows: 21990
    })
})
```

Let us take a look at the first two stories to get a flavor of the dataset.

```
print("--- Inspecting the first two stories ---")

story_1 = raw_datasets['train'][0]['text']

print(f"\n--- Story Example 1   ---")
print(story_1)
```

```
story_2 = raw_datasets['train'][1]['text']

print(f"\n--- Story Example 2   ---")
print(story_2)
```

--- Inspecting the first two stories ---

--- Story Example 1   ---
One day, a little girl named Lily found a needle in her room. She knew it was difficult to play with it because it was sharp. Lily wanted to share the needle with her mom, so she could sew a button on her shirt.

Lily went to her mom and said, "Mom, I found this needle. Can you share it with me and sew my shirt?" Her mom smiled and said, "Yes, Lily, we can share the needle and fix your shirt."

Together, they shared the needle and sewed the button on Lily's shirt. It was not difficult for them because they were sharing and helping each other. After they finished, Lily thanked her mom for sharing the needle and fixing her shirt. They both felt happy because they had shared and worked together.

--- Story Example 2   ---
Once upon a time, there was a little car named Beep. Beep loved to go fast and play in the sun. Beep was a healthy car because he always had good fuel. Good fuel made Beep happy and strong.

One day, Beep was driving in the park when he saw a big tree. The tree had many leaves that were falling. Beep liked how the leaves fall and wanted to play with them. Beep drove under the tree and watched the leaves fall on him. He laughed and beeped his horn.

Beep played with the falling leaves all day. When it was time to go home, Beep knew he needed more fuel. He went to the fuel place and got more healthy fuel. Now, Beep was ready to go fast and play again the next day. And Beep lived happily ever after.

Train-Test-Validation Split

You may have noticed that the dataset has been provided with only the training and the validation splits; it is missing the test split. To be able to rigorously evaluate our model, we will create a held-out test split to assess its final performance on unseen data.

```
# --- Create the Train / Test Split ---
train_test_split = raw_datasets['train'].train_test_split(test_size=20000, seed=28)

# The output is a new DatasetDict.
dataset_splits = {
    'train': train_test_split['train'],
    'validation': raw_datasets['validation'],
    'test': train_test_split['test']
}

print("--- Dataset Splits ---")
print(dataset_splits)
```

Tokenization

For our initial model, we try character-level tokenization. This is a straightforward approach where every unique character found in the training data – "a", "b", "-", "!", and so on – is treated as a distinct token. The primary advantage of this method is its simplicity, which results in a small and manageable vocabulary. This simplicity in tokenization, however, places a larger learning burden on our model, as it must learn to spell words entirely from scratch.

```
# --- Tokenization ---
# We build the vocabulary ONLY from the training set.
all_text = " ".join(dataset_splits['train']['text'])
vocab = sorted(list(set(all_text)))
vocab_size = len(vocab)

stoi = { ch:i for i,ch in enumerate(vocab) }
itos = { i:ch for i,ch in enumerate(vocab) }
text_to_tokens = lambda text: [stoi[c] for c in text]
tokens_to_text = lambda tokens: ''.join([itos[i] for i in tokens])
```

CHAPTER 6   BUILDING A LANGUAGE MODEL FOR STORYTELLING

We first gather all text from the training split and, crucially, *only* the training split, to avoid any data leakage from our validation or test sets. We then identify every unique character to form the complete vocabulary. Based on this, we create two dictionary mappings: a string-to-integer (stoi) lookup and, conversely, an integer-to-string (itos) lookup. We use these mappings to construct our final text_to_tokens and tokens_to_text functions, which serve as translators between our textual stories and the numerical tensors our model understands.

Let us test our tokenization function with a small dummy example.

```
test_string = "Hello World!"

encoded_output = text_to_tokens(test_string)
decoded_output = tokens_to_text(encoded_output)

print(f"Original string:      '{test_string}'")
print(f"Encoded output:       {encoded_output}")
print(f"Reconstructed string: '{decoded_output}'")

assert test_string == decoded_output
print("\nSuccess!")
```

```
Original string:      'Hello World!'
Encoded output:       [42, 70, 77, 77, 80, 2, 57, 80, 83, 77, 69, 3]
Reconstructed string: 'Hello World!'
Success!
```

We will now tokenize our train, validation, and test data splits.

```
# Convert all splits into PyTorch Tensors
train_data = torch.tensor(text_to_tokens(" ".join(dataset_splits['train']['text'])), dtype=torch.long)
val_data = torch.tensor(text_to_tokens(" ".join(dataset_splits['validation']['text'])), dtype=torch.long)
test_data = torch.tensor(text_to_tokens(" ".join(dataset_splits['test']['text'])), dtype=torch.long)

print(f"Train tensor shape: {train_data.shape}")
print(f"Validation tensor shape: {val_data.shape}")
print(f"Test tensor shape: {test_data.shape}")
```

```
Train tensor shape: torch.Size([1884070091])
Validation tensor shape: torch.Size([19212307])
Test tensor shape: torch.Size([18022829])
```

We have now created long tensors out of each of the three data splits. To recap, we concatenated all the stories in the training data, separated by a single space " ", into one long string. We then tokenized each character within this string and created a huge tensor out of these tokens. We repeated the same for the validation as well as the test set.

Batching

We shall now write a function to sample a batch from our training corpus. In all our previous chapters, we have used PyTorch's `Dataset` and `DataLoader` for batch loading. However, in this project, we use the simpler custom `get_batch()` function.

This shift in approach is due to the massive scale of our dataset and to demonstrate the established practices for training large-scale language models. A single pass, or "epoch," over the entire TinyStories training set would require hundreds of thousands of gradient updates and could take hours to complete, which would be impractical for us. Instead, our get_batch() function treats the data as a vast resource to be sampled at random, allowing us to train for a fixed number of iterations and achieve good results in a reasonable amount of time.

```python
def sample_batch(split_name):
    """
    Generates a small, random batch of data (inputs and targets)
    from the specified data split.
    """
    if split_name == 'train':
        source_data = train_data
    elif split_name == 'val':
        source_data = val_data
    else: # split_name == 'test'
        source_data = test_data

    # Generate 'batch_size' number of random starting indices
    start_indices = torch.randint(len(source_data) - block_size,
    (batch_size,))
```

```python
    # Create the input sequences by stacking slices from data
    input_sequences = torch.stack([source_data[i : i + block_size] for i in
    start_indices])

    # Create the target sequences by stacking the shifted slices
    target_sequences = torch.stack([source_data[i + 1 : i + block_size + 1]
    for i in start_indices])

    # Move the data to the appropriate device (CPU or GPU)
    input_sequences, target_sequences = input_sequences.to(device), target_
    sequences.to(device)

    return input_sequences, target_sequences
# --- Let's test our batch generator ---
sample_inputs, sample_targets = sample_batch('train')

print("Input batch shape:", sample_inputs.shape)
print("Target batch shape:", sample_targets.shape)

print("\nExamining the first example in the sample batch:")
print("Input context (first 10 tokens): ", sample_inputs[0, :10].tolist())
print("Target (first 10 tokens):        ", sample_targets[0, :10].
tolist())

Input batch shape: torch.Size([64, 128])
Target batch shape: torch.Size([64, 128])

Examining the first example in the sample batch:
Input context (first 10 tokens): [72, 66, 74, 79, 14, 2, 35, 79, 79, 66]
Target (first 10 tokens):        [66, 74, 79, 14, 2, 35, 79, 79, 66, 16]
```

Each call to sample_batch randomly selects batch_size starting positions from the specified data split, e.g., "train". For every such starting point, it creates an input sequence of block_size subsequent tokens. It also generates a corresponding target sequence, which is simply the input sequence shifted one token to the right. You can observe in the printed example as well that the target is simply a token-shifted version of the input.

A Simple Baseline: Bigram Model

We start our language modeling journey with a very simple baseline model called the bigram model. As discussed in the introduction, the core idea of a bigram model is that it attempts to predict the next token based *only* on the immediately preceding token, ignoring all the tokens prior to that.

Model Architecture

Let us define our bigram model now.

```
class BigramModel(nn.Module):
    """
    A simple baseline model that predicts the next token based only on the
    current token. It uses an embedding table as a direct lookup
    for logits.
    """
    def __init__(self, vocabulary_size):
        super().__init__()
        self.embedding_table = nn.Embedding(vocabulary_size,
        vocabulary_size)

    def forward(self, input_tokens):
        """
        - input_tokens: (B, T) tensor of token indices.
        - returns: (B, T, C) tensor of logits, where C is vocabulary_size.
        """
        logits = self.embedding_table(input_tokens)
        return logits
```

Just like in our previous chapters, our BigramModel class inherits from torch.nn.Module, which is the base class for all neural network modules in PyTorch.

The model contains just an embedding table, which is initialized in the __init__ method and accessed in the forward method. As we saw in Chapter 4, nn.Embedding is essentially a lookup table, which in this case would be a matrix of size vocabulary_size × vocabulary_size. For any token in the vocabulary, it would return a vector of size vocabulary_size. We will interpret this output vector as logits, i.e., unnormalized prediction scores, for every possible next character in our vocabulary.

The forward method takes as an argument the input_tokens, which is a 2D tensor of shape (B, T), representing a batch of B sequences, each with T token indices. Calling self.embedding_table(input_tokens) performs a lookup for every single integer token in the input_tokens tensor, retrieving a vector of size vocabulary_size for each integer token. The result is a new 3D tensor, logits, with the shape (B, T, C), where C is our vocabulary_size.

To summarize, for every token in every sequence of our batch, our model has produced a list of scores predicting which character will come next. Because this prediction is based on a direct lookup and is completely independent of any preceding tokens, it perfectly implements the simple "bigram" logic.

Generate Text from Model

Before even training the model, we will write a function to generate text from the model, one token (i.e., character) at a time. We shall be using a process called **autoregressive generation**. This works by building a sequence one token at a time, where each newly generated token becomes part of the input for predicting the next one. This "self-feeding" loop is the fundamental principle behind all modern generative text models like GPT.

```
def generate_text(model, context_tokens, max_new_tokens, block_size):
    """
    Generates new tokens from a model
    """
    print(f"Initial context: {context_tokens.tolist()}, Shape: {context_tokens.shape}")
    print("="*50)

    # The `context_tokens` tensor will be expanded in the loop
    for i in range(max_new_tokens):
        print(f"\n>>> STEP {i+1} <<<")

        # 1. Crop the context
        context_cropped = context_tokens[:, -block_size:]
        print(f"1. Cropped context: {context_cropped.tolist()}, Shape: {context_cropped.shape}")
```

```
# 2. Get the logits from the model
logits = model(context_cropped)
print(f"2. Logits tensor shape from model: {logits.shape}")

# 3. Focus only on the logit for the last time step
last_step_logits = logits[:, -1, :]
print(f"3. Logits for the last step shape: {last_step_logits.shape}")

# 4. Apply softmax to get probabilities
probabilities = F.softmax(last_step_logits, dim=-1)
print(f"4. Probabilities tensor shape: {probabilities.shape}")

# 5. Sample the next token from the probability distribution
next_token_index = torch.multinomial(probabilities, num_samples=1)
print(f"5. Sampled next token index: {next_token_index}, Shape: {next_token_index.shape}")
print(f"   (Sampled character: '{tokens_to_text(next_token_index.tolist()[0])}')")

# 6. Append the new token to our running sequence
context_tokens = torch.cat((context_tokens, next_token_index), dim=1)
print(f"6. New context sequence: {context_tokens.tolist()}, Shape: {context_tokens.shape}")
print("="*50)

return context_tokens
```

The above function involves several PyTorch gymnastics, and to help the reader understand these quickly, we have printed out the shapes of the tensors involved at each step. Later, when we run this during the actual model training and evaluation, we will run it after removing all the print statements, and we recommend that the reader do the same.

We shall now call the above function with a small dummy prompt. Also, we use a block_size of 4 just for demonstration purposes. We will walk through the function step by step with the help of this dummy test run.

```
bigram_model = BigramModel(vocab_size)
model_on_device = bigram_model.to(device)

prompt = "One day"
initial_context = torch.tensor([text_to_tokens(prompt)], dtype=torch.long,
device=device)

# Generate just 3 new tokens to keep the output simple
generate_text(
    model=model_on_device,
    context_tokens=initial_context,
    max_new_tokens=3,
    block_size=4
)
```

This prints out the following:

```
Initial context: [[49, 79, 70, 2, 69, 66, 90]], Shape: torch.Size([1, 7])
===========================================================================
>>> STEP 1 <<<
1. Cropped context: [[2, 69, 66, 90]], Shape: torch.Size([1, 4])
2. Logits tensor shape from model: torch.Size([1, 4, 174])
3. Logits for the last step shape: torch.Size([1, 174])
4. Probabilities tensor shape: torch.Size([1, 174])
5. Sampled next token index: tensor([[49]]), Shape: torch.Size([1, 1])
   (Sampled character: 'O')
6. New context sequence: [[49, 79, 70, 2, 69, 66, 90, 49]], Shape: torch.
   Size([1, 8])
===========================================================================
>>> STEP 2 <<<
1. Cropped context: [[69, 66, 90, 49]], Shape: torch.Size([1, 4])
2. Logits tensor shape from model: torch.Size([1, 4, 174])
3. Logits for the last step shape: torch.Size([1, 174])
4. Probabilities tensor shape: torch.Size([1, 174])
5. Sampled next token index: tensor([[12]]), Shape: torch.Size([1, 1])
   (Sampled character: '*')
```

CHAPTER 6   BUILDING A LANGUAGE MODEL FOR STORYTELLING

6. New context sequence: [[49, 79, 70, 2, 69, 66, 90, 49, 12]], Shape: torch.Size([1, 9])

==

>>> STEP 3 <<<
1. Cropped context: [[66, 90, 49, 12]], Shape: torch.Size([1, 4])
2. Logits tensor shape from model: torch.Size([1, 4, 174])
3. Logits for the last step shape: torch.Size([1, 174])
4. Probabilities tensor shape: torch.Size([1, 174])
5. Sampled next token index: tensor([[66]]), Shape: torch.Size([1, 1]) (Sampled character: 'a')
6. New context sequence: [[49, 79, 70, 2, 69, 66, 90, 49, 12, 66]], Shape: torch.Size([1, 10])

==

tensor([[49, 79, 70, 2, 69, 66, 90, 49, 12, 66]])

Initial Context: The function begins with our prompt "One day", a tensor of shape [1, 7].

Step 1:

1. **Cropping the Context:** The function first ensures the input to the model is no longer than its `block_size`. Our seven-token prompt is cropped to the last four tokens, resulting in a [1, 4] tensor.

2. **Getting Model Predictions:** This [1, 4] tensor is fed into the model. The model outputs a `logits` tensor of shape [1, 4, 174], meaning it has produced a vector of 174 prediction scores, which is the size of our vocabulary, for each of the four input tokens.

3. **Focusing on the Final Prediction:** We only care about what comes after the *last* token of our context. The code isolates the predictions for this final step, resulting in a tensor of shape [1, 174].

4. **Converting Logits to Probabilities:** These 174 scores are converted to probabilities using `softmax`.

5. **Sampling a New Token:** From this probability distribution, we sample one token. In this step, the model chose token 49 (the character "O").

6. **Appending the New Token:** Finally, this new token (49) is appended to our original seven-token sequence. As shown in line 6, our context has now grown to eight tokens long, ready for the next step.

Steps 2 and 3

The process now repeats itself. In step 2, the new eight-token context is cropped to the latest four tokens, fed to the model, and this time it samples token 12 (the character "*"). This is appended to our sequence, which now grows to nine tokens. This cycle of **crop-predict-sample-append** continues, as shown again in step 3, where the character "a" is generated.

This illustrates our process of autoregressive generation, where the model builds upon its own previous predictions, allowing it to write out text of any length, one character at a time.

Evaluate Function

We now write a function to evaluate any given model. We will use this function to evaluate our bigram model and then reuse it when we work on our attention-based model.

```python
def evaluate_model(model, sample_batch_fn, device, eval_iters, split):
    """
    Evaluates the model's performance on a given data split,
    calculating loss and perplexity.
    """
    model.eval() # Set model to evaluation mode

    running_loss = 0.0
    total_samples = 0
    start_time = time.time()

    with torch.no_grad():
        for _ in range(eval_iters):
            # We get a batch of data using our custom function
            X, Y = sample_batch_fn(split)

            # Note: In language models, we often count performance
            per token
```

```
        num_tokens_in_batch = X.numel() # B * T

        logits = model(X)

        # Calculate loss, ensuring tensors are shaped correctly
        B, T, C = logits.shape
        loss = F.cross_entropy(logits.view(B*T, C), Y.view(B*T))

        # Accumulate loss, weighted by the number of tokens in
        the batch
        running_loss += loss.item() * num_tokens_in_batch
        total_samples += num_tokens_in_batch
    end_time = time.time()

    # Handle potential division by zero if eval_iters is 0
    if total_samples == 0:
        print(f"Warning: No samples evaluated for {split} split.")
        return 0.0, float('inf')

    # Calculate final metrics
    eval_loss = running_loss / total_samples
    eval_perplexity = math.exp(eval_loss) # Perplexity is e^(loss)
    eval_duration = end_time - start_time

    # Print in your desired format
    print(f"{split} Loss: {eval_loss:.4f} | {split} Perplexity: {eval_
    perplexity:.2f} | Duration: {eval_duration:.2f}s")

    model.train() # Set model back to training mode
    return eval_loss, eval_perplexity
```

In this function, we essentially loop for a specified number of iterations and sample a batch from the provided data split in each iteration. Recall that the input X contains a batch of B, i.e., batch_size, samples, where each sample contains a sequence of T, i.e., block_size, tokens.

We then pass these inputs X to the model to get the model's raw prediction scores, i.e., logits.

CHAPTER 6 BUILDING A LANGUAGE MODEL FOR STORYTELLING

The model returns logits for each of these B * T tokens, which are then compared to the target tokens, which are again a tensor of shape [B, T], to compute the cross-entropy loss.

Since the cross_entropy function computes the average cross-entropy across these B*T values, we multiply it back by the total number of tokens to get the total running loss. After completing all our iterations, we take the average of this running loss over all our samples.

We also compute the **perplexity**, which is the exponential of the cross-entropy loss. Perplexity provides a more intuitive measure of uncertainty; for example, a value of 5 means the model is, on average, as confused as if it were guessing between five equally likely tokens. So a smaller number corresponds to a more confident model.

Model Training

We first define our hyperparameters to be used for training.

```
LEARNING_RATE = 1e-3
MAX_ITERATIONS = 15000
EVAL_INTERVAL = 500
EVAL_ITERS = 200
BLOCK_SIZE = 128
```

We then define our training loop.

```
# --- Model and Optimizer Instantiation ---
model = BigramModel(vocab_size)
model_on_device = model.to(device)

# Print the number of parameters to see the model's scale
num_params = sum(p.numel() for p in model_on_device.parameters())
print(f"Model has {num_params/1e6:.2f}M parameters.")

# Create the AdamW optimizer
optimizer = torch.optim.AdamW(model_on_device.parameters(),
lr=LEARNING_RATE)

print("\n--- Starting Training ---")
for step in range(MAX_ITERATIONS):
```

```python
    # Periodically evaluate the model
    if step % EVAL_INTERVAL == 0 or step == MAX_ITERATIONS - 1:
        print(f"\n--- Evaluating at step {step} ---")
        # Use our custom evaluation function
        train_loss, train_ppl = evaluate_model(model_on_device, sample_
        batch, device, EVAL_ITERS, split='train')
        val_loss, val_ppl = evaluate_model(model_on_device, sample_batch,
        device, EVAL_ITERS, split='val')
        print("-" * 50)

    # Fetch a batch of data
    input_batch, target_batch = sample_batch('train')

    # Perform a forward pass to get the logits
    logits = model_on_device(input_batch)

    # Calculate the loss explicitly
    loss = F.cross_entropy(logits.view(-1, vocab_size), target_batch.
    view(-1))

    # Perform the backward pass and update the weights
    optimizer.zero_grad(set_to_none=True)
    loss.backward()
    optimizer.step()

print(f"\n--- Finished Training after {MAX_ITERATIONS} steps ---")
```

The training loop has several overlapping structures with our evaluation function. We loop for a specified number of iterations, and in each iteration, we sample a batch randomly from the training data. We pass this sample batch to the model to get raw model prediction logits. These are then used to compute the cross-entropy loss, followed by the backpropagation step orchestrated by loss.backward() and optimizer.step(), as you must be familiar with by now.

Note that PyTorch's cross-entropy function is designed to work on a simple list of predictions. It expects a 2D tensor of shape (N, C), where N is the total number of items to classify and C is the number of classes. Therefore, we reshape its input using the .view method. The shape of logits is [B, T, C] to begin with. In the .view() method, we explicitly set the second dimension's size to be our vocab_size, which corresponds to C.

The first argument is set to -1; this is a special placeholder in PyTorch, instructing it to automatically infer the correct size for this dimension based on the total number of elements in the original tensor, i.e., B * T.

Model Testing

We test our model on the held-out test data to get a fair evaluation of our model.

```
test_loss, test_perplexity = evaluate_model(
    model=model_on_device,
    sample_batch_fn=sample_batch,
    device=device,
    eval_iters=500,
    split='test'
)

print("=" * 50)
print(f"Final Test Results for Bigram Model:")
print(f"  -> Test Loss: {test_loss:.4f}")
print(f"  -> Test Perplexity: {test_perplexity:.2f}")
print("=" * 50)
test Loss: 2.2992 | test Perplexity: 9.97 | Duration: 1.78s
========================================================================
Final Test Results for Bigram Model:
  -> Test Loss: 2.2992
  -> Test Perplexity: 9.97
========================================================================
```

Model Text Generation

We now use our trained bigram model to generate new text using our generate_text function.

```
# --- Final Generation ---
print("\n--- Generating Text from Trained Bigram Model ---")
initial_context = torch.zeros((1, 1), dtype=torch.long, device=device)
```

```
generated_tokens = generate_text(model=bigram_model,
                                 context_tokens=initial_context,
                                 max_new_tokens=500,
                                 block_size=BLOCK_SIZE)
print(tokens_to_text(generated_tokens[0].tolist()))
```

The text generated by this simple model is completely gibberish, unfortunately, with only a few coherent words such as "That", "are", "at", etc. Note that since we used character-level encoding, the model had to even learn how to spell words out, so having correctly spelled words in the text is an achievement, albeit falling short of our bar.

```
å©I¤Ast id l soma rt hes amu he d That'save t was t brd are ane b. Hino vee
"And bise toun. d thacin ay agreenckin bitert boy ceeved ave at Shedaso abe
tow. nd eireie thasho!"
Tome ainck stthed I aitind. sk bud pithiloramed san he lwas flor ied
toireask c"He iforitl. " "Lino t nd I hansalas. the whe Sut, unay s ary is
om. gl ay, andoconhesaloy athed t spey alacla she he Shet k."

Ond y sad hiey nd fome He ane fugs woutis ked tlyo t rutade. angange s en
Jay t apout hed wor. wad owar fll s bed res
```

Decoder-Only Transformer Model

We shall now move to a more interesting model, which builds a language model based on the decoder part of the classic transformer model.

Model Config

We define a model config object that can be passed to the Model's init function. This helps define all the parameters of the model architecture, such as the number of heads, etc.

```
from dataclasses import dataclass

@dataclass
class ModelConfig:
    """A class to hold the configuration for our language model."""
    block_size: int
```

```
    vocab_size: int
    n_layer: int
    n_head: int
    n_embd: int
    dropout: float = 0.2 # Default dropout rate
```

The @dataclass decorator we use for our ModelConfig class is a feature from Python's standard library, designed to reduce boilerplate code. Its main purpose is to automatically generate standard "dunder" (double-underscore) methods for classes that are primarily used for storing data. By simply adding the decorator and declaring our fields with type information, Python automatically writes essential methods for us, such as an __init__ method to assign arguments to instances and a __repr__ method to provide a clean, readable printout of the object and its values when debugging.

We now instantiate the ModelConfig for our transformer model.

```
config = ModelConfig(
    block_size=128,
    vocab_size=vocab_size,
    n_layer=6,
    n_head=6,
    n_embd=384,
    dropout=0.2
)
print(config)

>>> ModelConfig(block_size=128, vocab_size=174, n_layer=6, n_head=6, n_embd=384, dropout=0.2)
```

As you can see from the printed output, all the data members, such as block_size, etc., have been correctly instantiated, and are printed out neatly by the "print" function. This implies that both __init__ and __repr__ have been defined appropriately, thanks to the @dataclass decorator.

Model Architecture

For ease of instruction and to keep the code modular, we shall have four classes, where each class will utilize the preceding one: AttentionHead, MultiheadedAttention, DecoderBlock, and LanguageModel.

Attention Head

```python
class AttentionHead(nn.Module):
    """ A single head of self-attention """

    def __init__(self, config, head_size):
        super().__init__()
        # Linear layers to project the input into Query, Key, and
        Value spaces
        self.query_projection = nn.Linear(config.n_embd, head_size,
        bias=False)
        self.key_projection = nn.Linear(config.n_embd, head_size,
        bias=False)
        self.value_projection = nn.Linear(config.n_embd, head_size,
        bias=False)

        self.register_buffer('tril', torch.tril(torch.ones(config.block_
        size, config.block_size)))
        self.dropout = nn.Dropout(config.dropout)
        self.head_size = head_size

    def forward(self, x):
        B, T, C = x.shape

        # Project x to get the query, key, and value tensors
        query = self.query_projection(x) # (B, T, head_size)
        key = self.key_projection(x)     # (B, T, head_size)
        value = self.value_projection(x) # (B, T, head_size)

        # Compute attention scores

        # 1. Transpose the key tensor for matrix multiplication
        key_transposed = key.transpose(-2, -1) # (B, C, T)

        # 2. Compute the raw attention scores
        raw_attention_scores = torch.matmul(query, key_transposed) #
        (B, T, T)

        # 3. Scale the scores to prevent softmax saturation
        attention_scores = raw_attention_scores / math.sqrt(self.head_size)
```

```
attention_scores = attention_scores.masked_fill(self.tril[:T, :T]
== 0, float('-inf'))
attention_weights = F.softmax(attention_scores, dim=-1)
attention_weights = self.dropout(attention_weights)

# Perform the weighted aggregation of the values
out = torch.matmul(attention_weights, value)
return out
```

Let us go over the **__init__()** **method** first. The first three lines are the query, key, and value projection layers, which are all linear layers projecting from dimension n_embd to head_size. The next line deserves further scrutiny:

```
self.register_buffer('tril', torch.tril(torch.ones(config.block_size,
config.block_size)))
```

This is responsible for creating and registering our causal mask, which is essential for ensuring the model cannot "peek ahead" at future tokens during training. Working from the inside out, torch.ones(config.block_size, config.block_size)) simply creates a 2D tensor, which is like a square matrix, of dimension config.block_size by config.block_size. Torch's tril function creates a lower-triangular matrix containing 1s on and below the main diagonal and zeros above the main diagonal. This tensor represents our causal mask.

While we could assign this mask directly (e.g., self.tril = torch.tril(torch.ones(…))), using the register_buffer() method is the correct approach as it formally integrates the tensor into the model's state. It stipulates that the tril variable should be part of the model's persistent state but is not a learnable parameter. This provides two critical benefits: the tril tensor is automatically moved to the correct device (e.g., a GPU) when we call model.to(device), and it is properly saved and loaded with the model's state dictionary. Unlike an nn.Parameter, a buffer is ignored by the optimizer, ensuring it is never trained.

The **forward** method executes the following steps:

1. **Linear Projections:** First, the method uses three distinct linear layers – query_projection, key_projection, value_projection – to transform the input token representations. You can think of this as assigning roles: each token's vector is projected into three

specialized vectors – a **query** (what I'm looking for), a **key** (what information I have), and a **value** (the actual content I'll share if there's a match).

2. **Scaled Attention Scores:** Next, the model computes the raw attention scores to determine how much each token should interact with every other token. This happens in three steps:

   a. Transposing the *key* tensors in the last two dimensions

   b. Performing matrix multiplication over the query and the key_transposed matrices

   c. Normalizing by the square root of the last dimension of the key

3. **Causal Mask:** Next, we apply the causal mask; we first select the top T rows and columns of the mask, i.e., self.tril, and set the value of our attention_scores to negative infinity wherever the mask has a zero value.

4. **Softmax:** We then apply softmax over the last dimension (dim = -1), which essentially ensures that each row of the attention scores matrix corresponds to a probability distribution, with values between 0 and 1, which sum to 1.

5. **Dropout:** Thereafter, we apply dropout to these weights, which randomly sets some weights to zero during training to prevent the model from overfitting.

6. **Weighing the Values:** In the final step, torch.matmul(attention_weights, value) is performed. This takes the weighted sum of the value vectors of all other tokens in the context, weighted by the attention_weights.

Multi-headed Attention

```
class MultiHeadAttention(nn.Module):
    """ Multiple heads of self-attention in parallel """

    def __init__(self, config: ModelConfig):
        super().__init__()
```

```python
        # Ensure the embedding dimension is divisible by the number
        of heads
        assert config.n_embd % config.n_head == 0
        head_size = config.n_embd // config.n_head

        # A ModuleList to hold all the parallel attention heads
        self.attention_heads = nn.ModuleList([AttentionHead(config, head_
        size) for _ in range(config.n_head)])

        # A final linear layer to project the concatenated outputs
        self.output_projection = nn.Linear(config.n_embd, config.n_embd)
        self.dropout = nn.Dropout(config.dropout)

    def forward(self, x):
        # Run each attention head in parallel and concatenate their outputs
        concatenated_heads = torch.cat([head(x) for head in self.attention_
        heads], dim=-1)

        # Apply the final projection and dropout
        projected_output = self.dropout(self.output_
        projection(concatenated_heads))

        return projected_output
```

Let us first examine what the __init__() method does.
The constructor sets up the layers needed for multi-head attention.

1. It starts with a sanity check to ensure that the total embedding dimension can be split evenly among the number of heads we want.

2. It then calculates the dimension for each individual head by dividing the total embedding dimension size by the number of heads.

3. We then use nn.ModuleList to hold our list of attention heads. nn.ModuleList is a special PyTorch list that properly registers the parameters of all the modules within that list with our model. Note that had we stored our attention heads in a standard Python

list, their parameters would not have been registered with the model, and consequently, the optimizer wouldn't have updated them during training.

4. We then add a final, learnable linear layer. Its main job is to combine the outputs from all the parallel heads and learn how to best blend them back into a single, cohesive representation.

The forward method is now fairly straightforward: it passes the input through all the attention heads in parallel (i.e., all the heads get the same input x). Each head produces its own output tensor of shape (B, T, head_size). Torch.cat then concatenates all these output tensors, resulting in a single tensor of shape (B, T, n_embd), since we had head_size * n_head = n_embd. This output is then passed through a final output projection layer.

Decoder Block

We now define the decoder block, which combines the MultiHeadAttention block we just defined, with a simple feed-forward network comprising two linear layers with a ReLU activation and dropout, some layer normalizations, and residual connections.

```
class DecoderBlock(nn.Module):
    """
    A Transformer Decoder Block, which combines self-attention
    and a feed-forward network with residual connections and layer
    normalization.
    """
    def __init__(self, config: ModelConfig):
        super().__init__()

        self.self_attention = MultiHeadAttention(config)
        self.feed_forward = nn.Sequential(
            nn.Linear(config.n_embd, 4 * config.n_embd),
            nn.ReLU(),
            nn.Linear(4 * config.n_embd, config.n_embd),
            nn.Dropout(config.dropout)
        )
```

```
        self.layer_norm1 = nn.LayerNorm(config.n_embd)
        self.layer_norm2 = nn.LayerNorm(config.n_embd)

    def forward(self, x):
        """
        Defines the data flow through the block
        This follows the "Pre-LN" (Layer Normalization first) architecture.
        """
        # --- First sub-layer: Multi-Head Self-Attention ---
        attention_output = self.self_attention(self.layer_norm1(x))
        x = x + attention_output

        # --- Second sub-layer: Feed-Forward Network ---
        feed_forward_output = self.feed_forward(self.layer_norm2(x))
        x = x + feed_forward_output

        return x
```

The init function defines all the layers of the decoder architecture. The forward function applies these layers in the appropriate order. It implements the residual connections by adding x back to the application of the self_attention layer to x and similarly to the feed_forward layer's output.

This decoder block is the repeatable, building block of our Transformer architecture.

Language Model

We define our final language model within its namesake class, LanguageModel.

```
class LanguageModel(nn.Module):
    def __init__(self, config):
        super().__init__()
        self.config = config

        # --- Layer Definitions ---
        self.token_embeddings = nn.Embedding(config.vocab_size,
        config.n_embd)
        self.positional_embeddings = nn.Embedding(config.block_size,
        config.n_embd)
```

CHAPTER 6   BUILDING A LANGUAGE MODEL FOR STORYTELLING

```python
        self.decoder_blocks = nn.Sequential(*[DecoderBlock(config) for _ in
        range(config.n_layer)])
        self.final_layer_norm = nn.LayerNorm(config.n_embd)
        self.logits_head = nn.Linear(config.n_embd, config.vocab_size)

    def forward(self, token_indices: torch.Tensor):
        """
        Performs a forward pass, returning the raw logits.
        - token_indices: (B, T) tensor of token indices.
        """
        batch_size, seq_len = token_indices.shape

        # Get token and position embeddings
        token_vecs = self.token_embeddings(token_indices) # (B, T, n_embd)
        position_vecs = self.positional_embeddings(torch.arange(seq_len,
        device=token_indices.device)) # (T, n_embd)

        # Add them together and pass through the Transformer blocks
        x = token_vecs + position_vecs # Broadcasting adds position_vecs to
        each sequence in the batch
        x = self.decoder_blocks(x)
        x = self.final_layer_norm(x)

        # Get the final logits
        logits = self.logits_head(x) # (B, T, vocab_size)

        return logits
```

The init method creates two distinct embedding tables: one to represent the meaning of each token (token_embeddings) and another to represent the position of each token in the sequence (positional_embeddings). The core of the model is the stack of our previously defined DecoderBlocks of size n_layer defined in the config. These are stored in an nn.Sequential container. Finally, it defines a concluding LayerNorm and a final linear layer (logits_head) to produce the output predictions.

The forward method defines the explicit path that the data takes through the model's layers, defined in the init function.

Model Training

We can reuse the exact same code for the training loop as the one we wrote for our bigram model. Only the hyperparameters and the model setup need to change, as

```
LEARNING_RATE = 3e-4
MAX_ITERATIONS = 15000

EVAL_INTERVAL = 500
EVAL_ITERS = 200

config = ModelConfig(
    block_size=block_size,
    vocab_size=vocab_size,
    n_layer=6,
    n_head=6,
    n_embd=384,
    dropout=0.2
)
model = LanguageModel(config)
```

The rest of the code can be copied exactly from the bigram model training section.

Model Testing

The testing code we wrote for the bigram model can be used as is for evaluating our transformer model. It gave me the following results on my machine:

```
test Loss: 0.7593 | test Perplexity: 2.14 | Duration: 10.15s
```

We reproduce the results for our bigram model here for comparison.

```
test Loss: 2.2992 | test Perplexity: 9.97 | Duration: 1.78s
```

As you can see, the test loss and perplexity are significantly better for the transformer model. It takes much longer for inference, however.

Text Generation

Again, using the same code as for the bigram model for text generation yielded the following passage using our transformer-based language model:

aid means what it's hard and fun on the ground. Lily thought it was her favorite mom and she should fix the water. Once upon a time, there was a little girl named Lily. Tom was so excited because it made her mom every day to make her job. One day, Lily went to the water and addrashed valley. They walked to the backyard and placked their trunk. Lily saw a sape and manage about the flowers. She said, "Oh, warn do not to borrow a red phone. You need their dolphine and toys free. You need to be here

As you can notice, this now spells almost all the words correctly, and even the sentences are grammatically sound for the most part. However, they don't make perfect sense semantically, and don't sound too coherent. This is still a significant improvement over the bigram model, which hardly produced any sensible words, let alone sentences.

Decoder-Only Model with Word-Level Tokenization

We have been careful to make our code quite modular, which makes it straightforward to switch from character-level tokenization to word-level tokenization. It only needs changes in the code related to tokenization:

```
import collections
# --- 1. Build the Vocabulary ---
words = " ".join(dataset_splits['train']['text']).split()

# Build a frequency counter for all words
word_counts = collections.Counter(words)

# Filter to only keep words that appear more than a threshold
MIN_FREQUENCY = 10
vocab = sorted([word for word, count in word_counts.items() if count >= MIN_FREQUENCY])

# Add a special token for "unknown" words.
vocab.insert(0, '<unk>')
vocab_size = len(vocab)
```

```
print(f"--- Word-Level Vocabulary ---")
print(f"Total unique words before filtering: {len(word_counts)}")
print(f"New vocabulary size (min freq >= {MIN_FREQUENCY}): {vocab_size}")
print(f"First 100 words in vocab: {vocab[:100]}")

# --- 2. Create the Tokenizer Mappings ---
stoi = { word:i for i,word in enumerate(vocab) }
itos = { i:word for i,word in enumerate(vocab) }
unknown_token_id = stoi['<unk>']

# --- 3. Create the Tokenizer and De-tokenizer Functions ---
text_to_tokens = lambda text: [stoi.get(word, unknown_token_id) for word in text.split()]
tokens_to_text = lambda tokens: ' '.join([itos[i] for i in tokens])

# --- 4. Test the New Tokenizer ---
test_string = "Once upon a time, there was a boy named Timmy."
encoded_output = text_to_tokens(test_string)
decoded_output = tokens_to_text(encoded_output)

print("\n--- Word-Level Tokenizer Test ---")
print(f"Original string:       '{test_string}'")
print(f"Tokenized output:      {encoded_output}")
print(f"Reconstructed string: '{decoded_output}'")
--- Word-Level Vocabulary ---
Total unique words before filtering: 264946
New vocabulary size (min freq >= 10): 76504
```

Notice that the vocabulary now is 76K, significantly larger than that during character-level tokenization, when it was merely 174. For this reason, we use a slightly larger model now, with the following config:

```
config = ModelConfig(
    block_size=128,
    vocab_size=vocab_size,
    n_layer=6,
    n_head=6,
```

```
    n_embd=384,
    dropout=0.2
)
```

This model now has 69M parameters! After training it for 15K steps, it gave me a validation loss of 2.469, with a perplexity of 11.81. It is tempting to compare this with the perplexity of 2.14 we had achieved during the character-level tokenization. But this would be an unfair comparison, because for character-level tokenization, it signifies that the model is confident about 2.1 tokens from among 174 candidate tokens. In contrast, in our word-level tokenization example, the model has narrowed down to 11.81 tokens from among 76K candidate tokens, which is a much bigger achievement.

This new model generates the following story.

```
<unk> and as he lifted the string on a big rock, he tried to tie the string
with his foot. His foot slipped from it, because he wanted to get rid of
it. But his deep ankle was so weak that he wouldn't listen. He kept trying,
slipping around until he was about to shoot out of the branch. He soon
realized that the string was making a lovely valuable sound of something
every day. The sailboat was so excited and sparkly and danced around trying
to think of what it had to do. The mother decided to close the box
```

This still makes little sense as a story. But the sentences still seem much more coherent now than before.

Further Improvements

As next steps, we would recommend the following improvements to get even more coherent stories:

1. **Increase the Block Size**: Currently, the model is exposed to a relatively short context window of 128 tokens. It does not remember anything beyond this window size. Increasing this to 256 or 512 will allow the model to remember better about the context in which it should build the story.

2. **Increase the Model Capacity**: Increasing the size of different model parameters, such as the number of heads, embedding dimension, etc., will only help make the model more complex, which will enable it to learn more nuances in storytelling.

3. **Train for Longer**: Since the above two improvements aim at making the model more complex, you will need to train the model for more iterations for it to reach its potential.

4. **Upgrade to a Subword Tokenizer:** Professional models often use a **subword tokenizer**. This method breaks rare words down into meaningful parts (e.g., "tokenization" -> ["token", "ization"]) while keeping common words as single units. This provides the best of both worlds: a manageable vocabulary size and the ability to understand word structure, which dramatically improves the model's semantic understanding.

All these steps require using heavier computing resources for longer time periods, but they are sure to improve the end results.

Troubleshooting

While training large models, you may encounter practical issues. Here are some common issues and some possible fixes you can try:

- **Out of Memory:** This might happen if a prior model (like the bigram model) is still on the device while you try to load the next one. To fix this, restart the runtime before loading the second model. If the error persists, try reducing the batch_size, block_size, or n_embd.

- **Diverging Loss:** Lower the learning rate (e.g., from 3e-4 to 1e-4), enable gradient clipping, and confirm your labels are long dtype.

Summary

- **Language modeling** is the foundational NLP task of assigning a probability to the next word in a sequence. Practically, it involves training a model to predict the next word or token in a sequence.

- Historically, statistical methods such as **n-gram models** were the most popular approach toward language modeling. N-gram models are based on the idea of estimating the probability of the next word based on only the preceding $n - 1$ words.

- Further advancements in this area were brought **by recurrent neural networks (RNNs) and long short-term memory (LSTMs)**, which introduced a form of memory but were limited by their sequential nature.

- The **transformer** architecture marked a significant advancement, allowing for parallel computation and improved handling of long-range dependencies using the **self-attention mechanism**.

- Self-attention is based on projecting each token into a **query, key, and value vector**, allowing it to attend to other relevant tokens in its context.

- In this way, an **attention head** block refines the representation of a token using other relevant tokens in its context. The transformer uses **multi-head attention**, i.e., it employs multiple heads in each layer, where each head specializes in a different aspect of the data.

- A **decoder-only transformer** – which we use in our project – comprises a multi-head attention layer, a feed-forward layer, along with residual connections and layer normalizations.

- A common architecture for building a language model is obtained by stacking many such decoder blocks.

- **Positional embeddings** are constructed to convey to the model a word's position in the sequence. They are added to the token embeddings and passed to the model.

- We worked on the hands-on project aimed at building a **storytelling language model** trained on the `TinyStories` dataset using character-level tokenization.

- We started with a simple **bigram model baseline** that predicts the next token based only on the current one.

- We then upgraded to the full **decoder-only transformer**, implementing the modular components like AttentionHead, MultiHeadAttention, and DecoderBlock along with residual connections and LayerNorm.

- Text was generated using an **autoregressive process**, where the model's own previous output was used as the input for the next step.

- We used the **cross-entropy loss** for training and evaluated the model using both the loss and **perplexity**. The perplexity measure provides a more intuitive measure of how "confident" the model is with respect to the number of choices it's considering for the next token.

- The project used iteration-based training, where a fixed number of batches were randomly sampled from the massive text corpus; this is a modern paradigm better suited for large datasets than traditional epoch-based training.

- Finally, we switched from character-level tokenization to **word-level tokenization** to generate (relatively) more coherent stories.

CHAPTER 7

Audio Classification with PyTorch

This chapter will introduce you to the fascinating world of audio processing and classification using machine learning. We will embark on a journey that begins with the fundamentals of sound and goes on to build two complete, practical projects. We'll start by exploring the physics of sound on a high level – how vibrations create sound waves – and how computers "hear" by digitizing these waves into a numerical format. You will then learn a crucial technique in modern audio AI: transforming raw audio waveforms into Mel spectrograms, which are powerful, image-like representations that make sounds much easier for ML models to classify.

We will then dive into our first project, which is an end-to-end **speech command recognition** system. Using the Google Speech Commands dataset, we will train a model to recognize simple spoken words. Building on these skills, we will then proceed to our second project, which works on the more nuanced task of **Speech Emotion Recognition (SER)**. Using the RAVDESS (Ryerson Audio-Visual Database of Emotional Speech and Song) dataset, which features actors expressing various emotions through their speech, we will train a model to understand *how* something is said and not *what* is said.

Overall, the chapter will teach the theoretical fundamentals of audio processing, along with the practical side of some standard data processing techniques for audio data, building custom CNN-based models, and data augmentation techniques, and also introduce some other class-specific metrics commonly used in classification problems.

Understanding Sound

When an object – such as a guitar string – vibrates, it causes the air surrounding it to vibrate; this "sound wave" eventually reaches your ears as music. Every sound we hear is essentially just vibrations traveling through a medium, such as air, as pressure waves.

If the guitar string is plucked with greater force, the resulting vibrations have a **higher amplitude**, and the sound you hear is louder. A gentle pluck produces vibrations with a **lower amplitude**, resulting in a softer sound.

Similarly, plucking a string tuned to a higher note generates sound waves with a **higher frequency** – meaning the vibrations occur more rapidly – while plucking a string tuned to a lower note produces **lower-frequency** waves, where the vibrations are slower. **Amplitude** controls the loudness of the sound, while **frequency** determines its pitch. Figure 7-1 illustrates how the loudness is determined by the sound wave's amplitude, whereas the pitch is determined by its frequency.

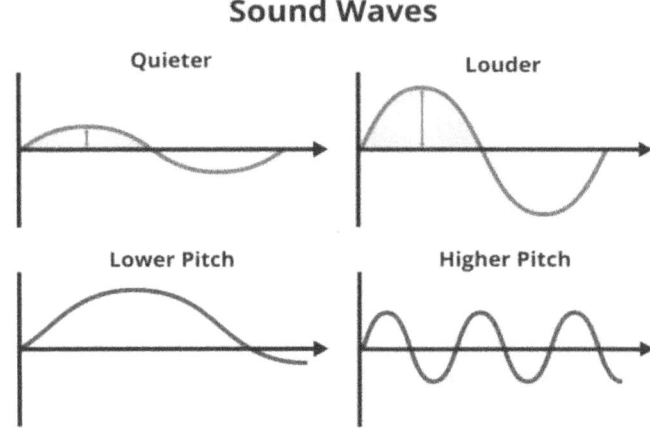

Figure 7-1. Sound wave (picture credit: istockphoto.com: https://www.istockphoto.com/vector/vector-graph-with-sound-waves-greater-amplitude-waves-mean-a-louder-sound-smaller-gm1421126126-466867167)

How Computers "Hear"

Just like computers see images as a matrix of numbers (pixels), they "hear" sound waves as a series of numbers. To capture this physical phenomenon of a continuous sound wave into a numerical series, we need to **sample** it, i.e., take snapshots of its amplitude at frequent, regular intervals. The number of snapshots we take per second is called its sampling rate or sampling frequency. For example, CD-quality audio uses 44,100 Hz (44.1 kHz), meaning 44,100 samples are taken every second. For speech, 16,000 Hz (16 kHz) is very common. You can think of it like frames per second in a video. Just like more frames per second give smoother motion, a higher sample rate captures the sound wave more accurately.

A key principle here is that to accurately represent a sound, our sample rate needs to be at least twice the highest frequency present in that sound. This is why capturing the full range of human hearing (up to ~20kHz) requires sample rates like 44.1kHz. This mathematically proven notion is called the Nyquist–Shannon theorem.

The amplitude of each of our samples needs to be represented with a finite-precision number. A process called **quantization** assigns each sample's amplitude to the closest level in a predefined set of discrete levels. The number of available levels is determined by the bit depth (e.g., 16-bit, 24-bit). More bits mean more levels, allowing for a more accurate representation of the amplitude. To understand this better, imagine measuring height with a ruler. A ruler with only centimeter markings (lower bit depth) is less precise than one with millimeter markings (higher bit depth).

After sampling and quantization, our original sound wave is now a sequence of numbers. This sequence is what we call the digital waveform – it's the primary way computers store and process audio. Typically, for mono (single-channel) audio, this is a 1D array of numbers. For stereo audio, it would be a 2D array (two channels).

Classifying Digital Waves

Consider the simple task of classifying sounds into buckets; for example, you have short clips of music, and you wish to classify each into its genre. Starting with these clips as digital waveforms, how would we go about classifying them using ML models? At first glance, it might seem straightforward – after all, each clip is already stored as a digital waveform, a series of amplitude values over time.

You might be tempted to feed these raw waveforms directly into an ML model and expect it to learn the patterns. However, this approach can quickly become challenging. The raw waveform is just a long stream of numbers, and without any structure or higher-level features, it can be challenging for a model to make sense of it.

Instead, a more effective and widely used technique is to transform the waveform into a **Mel spectrogram** – a two-dimensional, image-like representation that captures the sound's frequency content over time. This transformation, as we'll see in our upcoming projects, provides a much richer and more structured input for ML models to work with.

Mel Spectrogram: Turning Sounds into Model-Friendly Images (Optional)

Deconstructing a Waveform

When we look at a digital audio waveform, we see a faithful recording of how a sound's loudness changes from one moment to the next. But what it doesn't immediately show us is the sound's unique *character* or *timbre*. Is it the deep thud of a drum, a softly spoken word, or the distinct resonance of a violin string? These distinguishing qualities come from the rich and unique mix of different pitches, or **frequencies**, that make up the sound.

Hidden inside the single wavy line of a waveform is a combination of many simpler sound waves, all layered together. Nearly every sound we hear is a complex mixture of these individual waves, each oscillating at its own frequency and strength. To understand a sound's true nature, we need to break it down into these basic ingredients. For this, we use a powerful mathematical tool called the **Fourier transform**, which acts like a **prism for sound**. Just as a prism can take a single beam of white light and reveal the full spectrum of rainbow colors hidden within, the Fourier transform takes a complex waveform and reveals the full spectrum of simple frequencies that combine to create it. Figure 7-2 illustrates how the Fourier Transform decomposes a complex waveform into its constituent sine waves of different frequencies.

Figure 7-2. Source: ChatGPT

Moreover, this frequency recipe of a sound is a function of time. To capture this evolution, we use the **short-time Fourier transform (STFT)**. Instead of analyzing the entire audio clip at once, we slice it into short, overlapping windows (e.g., 25 milliseconds each) and perform a Fourier transform on each tiny window.

When we take the frequency recipe from each window and arrange them side-by-side in chronological order, we create a powerful visualization called a **spectrogram**. A spectrogram is a 2D image where

- The **horizontal axis (x axis)** is **time**.

- The **vertical axis (y axis)** is **frequency** (from low to high).

- The **color or brightness** at any point shows the **energy** or loudness of that frequency at that specific time. Brighter areas mean more energy.

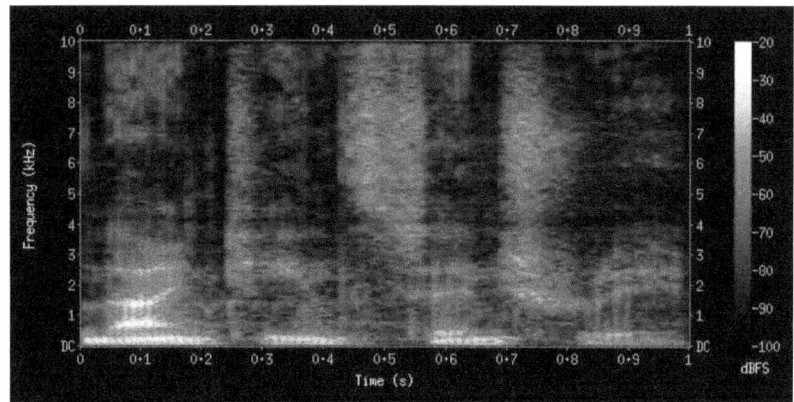

Figure 7-3. Spectrogram of the spoken words "nineteenth century". (Source: https://en.wikipedia.org/wiki/Spectrogram, By Aquegg - Own work, Public Domain, https://commons.wikimedia.org/w/index.php?curid=5544473)

This "image of sound" reveals the rich structure hidden in the waveform – the rising pitches, rhythmic beats, and rich harmonic textures – in a way we can visually interpret. As an example, Figure 7-3 depicts a spectrogram corresponding to the spoken words "nineteenth century".

The Mel Scale

The frequency axis of a standard spectrogram is linear, meaning the distance from 100 Hz to 200 Hz is the same as from 10,000 Hz to 10,100 Hz. This isn't how our perception of sound works. We are much more sensitive to changes in lower frequencies.

To make our feature more aligned with human perception, we use the **Mel scale**. It is a perceptual scale of pitches where equal distances on the scale sound equally distant to the human ear. It gives more resolution to lower frequencies and compresses the higher ones.

To create a **Mel spectrogram**, we take a standard power spectrogram and apply a set of **Mel filter banks**. Imagine these as triangular "listening windows" applied to the frequency axis. They are narrow and tightly packed at low frequencies but become wider and more spread out at higher frequencies, mimicking how our inner ear processes sound. Each filter bank sums up the energy it captures, resulting in a new frequency axis with Mel bins.

Furthermore, the energy in these Mel bins is converted to a logarithmic scale called **decibels (dB)**. This also mimics our perception of loudness, making small variations in quiet sounds more apparent while compressing the range of very loud sounds.

Mel Spectrograms for Classification

To summarize the above section, a Mel spectrogram translates complex audio into a structured, perceptually meaningful picture. We have now converted the challenge of audio classification into the familiar problem of image classification, a topic we explored in Chapter 3. We can therefore use the powerful tools of computer vision for our audio task. Specifically, we can now use convolutional neural networks (CNNs) to "see" and learn the distinctive patterns of speech or emotion or music, or any other aspect relevant to our specific task, within these spectrograms, an approach that will be the foundation for the projects that follow.

Project 1: Speech Command Recognition

In our first project, we'll dive into the fascinating world of **speech command recognition**. The goal is simple but impactful: given a short audio clip of someone speaking a single word – like "yes," "no," or "up" – we want to build an ML model that can correctly recognize which word was spoken.

To tackle this task, we'll use the **Speech Commands dataset**, a popular collection of thousands of short, one-second audio recordings of spoken words. It is perfect for our first project since it is clean, well-labeled, and easy to work with. Throughout this project, we'll learn how to load and process real audio data, convert it into a model-friendly format like a **Mel spectrogram**, and train a CNN model to classify the spoken commands.

Fun Fact: Speech command recognition is one of the core technologies behind voice-controlled devices like Alexa and Siri. Even though our project focuses on simple words, the principles are the same!

Imports

As always, we start with the necessary imports.

```
import torch
import torchaudio
import torchaudio.transforms as T
from torch.utils.data import DataLoader, TensorDataset
import os
import collections
from IPython.display import Audio # For playing audio in Jupyter
```

We shall be relying heavily on PyTorch's torchaudio library in this project.

Dataset

We work with the Speech Command dataset[1] created by Google specifically for this purpose. It's not designed for transcribing long, complex sentences, but rather to train models that can accurately recognize a small, predefined vocabulary of words, making it a perfect fit for our goal.

The dataset consists of over 100,000 short audio clips, each about one second long, contributed by thousands of different speakers. This diversity is key to building a model that can generalize to many different voices. The audio is provided in a standardized format (16kHz sample rate, mono channel), ready for processing. The dataset's structure is intuitive, with each audio clip stored in a folder named after the word being spoken.

The words themselves are divided into several conceptual categories, which is crucial for building a realistic system:

1. **Command Words:** This includes simple, direct commands like yes, no, up, down, left, right, on, off, stop, and go. These will be the primary keywords our model learns to identify.

2. **Auxiliary Words:** The dataset contains many other words, such as digits (zero, one, two…), and common nouns (bed, bird, cat, dog, tree,…). We will group all of these auxiliary words into a single **_unknown_** category. This teaches our model to explicitly

---

[1] https://arxiv.org/pdf/1804.03209

ignore words that are not part of our target command set – a critical feature for any real-world voice interface to prevent it from activating on irrelevant speech.

3. **Background Noise:** A special folder named _background_noise_ contains recordings of various ambient sounds like pink noise, people talking, and household machinery.

We will now load our dataset using the torchaudio library.

```
# 1. Define the root directory where the dataset will be downloaded and stored.
DATASET_ROOT = './speech_commands_data_download_test'

# Create the directory if it doesn't exist, although torchaudio often handles this.
os.makedirs(DATASET_ROOT, exist_ok=True)

print(f"Attempting to download and load Speech Commands dataset into: {DATASET_ROOT}")

train_data = torchaudio.datasets.SPEECHCOMMANDS(
    root=DATASET_ROOT,
    download=True,
    subset='training'
)
val_data = torchaudio.datasets.SPEECHCOMMANDS(
    root=DATASET_ROOT,
    download=True,
    subset='validation'
)
test_data = torchaudio.datasets.SPEECHCOMMANDS(
    root=DATASET_ROOT,
    download=True,
    subset='testing'
)
print(f"Dataset download and loading initiated.")
print(f"Number of items in the training subset: {len(train_data)}")
print(f"Number of items in the validation subset: {len(val_data)}")
print(f"Number of items in the testing subset: {len(test_data)}")
```

This prints out the following information about the data sizes:

```
Number of items in the training subset: 84843
Number of items in the validation subset: 9981
Number of items in the testing subset: 11005
```

Data Processing Parameters

We now define some hyperparameters for data processing and training.

```
batch_size = 64
target_sample_rate = 16000
num_samples_per_clip = 16000 # 1 second at 16kHz

n_fft = 400
hop_length = 160
n_mels = 64

device = torch.device("cuda" if torch.cuda.is_available() else "cpu")
print(f"Using device: {device}")
```

Here, we introduce the usage of the torch.device() function for the selection of the device to be used. If you have access to GPUs, say in an environment such as Google Colab, the device will show as cuda; if you don't have access to GPUs, this code should continue running smoothly on your CPUs.

EDA

Let us now look at an example from the dataset and see what it contains:

```
train_data[2000]
>>> (tensor([[-0.0012, -0.0013, -0.0014,  ...,  0.0007,  0.0013,  0.0010]]),
 16000,
 'bed',
 '69086eb0',
 0)
```

Each sample contains five values. The first is a tensor representing simply the waveform of the sample audio. Then we have the sample rate, which is 16000 for our sample, followed by the label, which is "bed" here. The fourth element is the speaker id, whereas the last value is the utterance number of that speaker of that utterance.

We can listen to any audio sample in the notebook itself:

```
sample_index = 10000
waveform, sample_rate, label, speaker_id, utterance_number = train_data[sample_index]

# Display the label so you know what word to expect
print(f"Playing audio for label: '{label}' (Sample Rate: {sample_rate} Hz)")

# Create an audio player widget in the Jupyter Notebook output
Audio(data=waveform.numpy(), rate=sample_rate)
```

We can also visualize the Mel spectrogram of this sample. Recall that we imported torch.transforms as T. We use the MelSpectrogram transform from this library, followed by the AmplitudeToDB transform. The AmplitudeToDB transform converts the spectrogram's linear energy values into the logarithmic decibel (dB) scale, which better reflects how humans perceive loudness, as we saw in the first section. We need to pass different parameters related to audio processing, such as sample_rate, n_fft, etc., to these transforms. Before these transforms, we may also need to resample the waveform if its original sample rate doesn't match our target sample rate. We achieve this via the Resample function from the transforms library.

```
mel_spectrogram_transform = T.MelSpectrogram(
    sample_rate=target_sample_rate,
    n_fft=n_fft,
    hop_length=hop_length,
    n_mels=n_mels,
    normalized=False
).to(device)

amplitude_to_db_transform = T.AmplitudeToDB(stype="power", top_db=80).to(device)
```

```
sample_index_to_plot = 20000 # Choose any sample
waveform, original_sr, label, _, _ = train_data[sample_index_to_plot]

print(f"Plotting for: '{label}', Original SR: {original_sr} Hz")

# 1. Resample to target_sample_rate
processed_waveform = waveform.to(device)
if original_sr != target_sample_rate:
    resampler = T.Resample(orig_freq=original_sr, new_freq=target_sample_
    rate).to(device)
    processed_waveform = resampler(processed_waveform)
    print(f"Resampled to {target_sample_rate} Hz for Mel spectrogram
    calculation.")

# 2. Apply Mel Spectrogram and DB conversion
mel_spec = mel_spectrogram_transform(processed_waveform)
mel_spec_db = amplitude_to_db_transform(mel_spec)

# 3. Prepare for plotting
plot_spectrogram_data = mel_spec_db.squeeze().cpu().numpy()

# 4. Plot
plt.figure(figsize=(10, 4))
plt.imshow(plot_spectrogram_data, aspect='auto', origin='lower',
cmap='viridis')
plt.colorbar(format='%+2.0f dB')
plt.title(f'Mel Spectrogram - Label: "{label}" (Actual Length)')
plt.xlabel('Time Frames')
plt.ylabel(f'Mel Bins (from SR: {target_sample_rate} Hz)')
plt.tight_layout()
plt.show()
```

This gives us the following Mel spectrogram (see Figure 7-4).

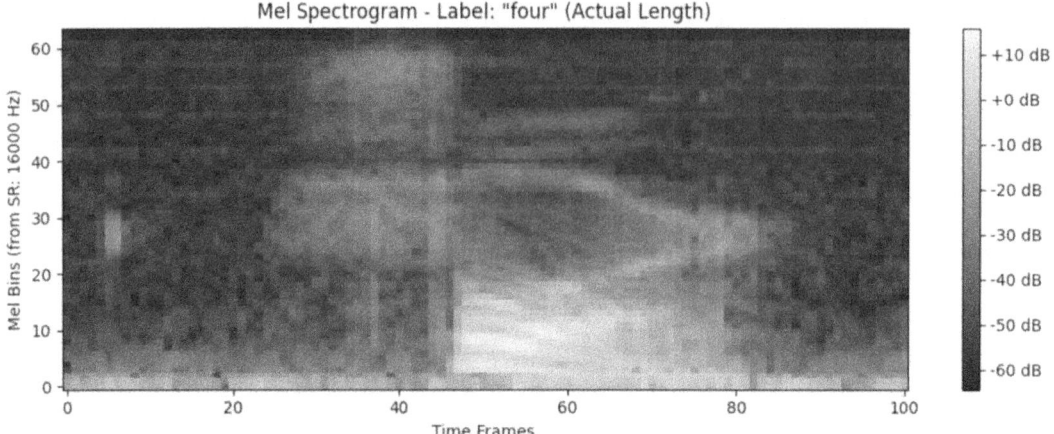

Figure 7-4. Sample Mel spectrogram

Mel Spectrogram Dimensions

For our sample, we calculate the dimensions of the Mel spectrogram transformation. The height of this is n_mels, i.e., the number of mel bins, which we had defined to be 64, whereas the width is the number of time frames, which we shall store as num_time_frames.

```
print(f"Sample waveform shape: {waveform.shape}")
print(f"Mel spectrogram output shape: {mel_spec.shape}")

# Extract the number of time frames and n_mels
num_time_frames = mel_spec.shape[-1]
n_mels = mel_spec.shape[-2]

print(f"Number of Mel bins (n_mels) from transform output: {n_mels}")
print(f"Number of time frames from transform output: {num_time_frames}")
```

This prints out the following:

```
Sample waveform shape: torch.Size([1, 16000])
Mel spectrogram output shape: torch.Size([1, 64, 101])
Number of Mel bins (n_mels) from transform output: 64
Number of time frames from transform output: 101
```

Note that we can also calculate the number of time frames by the analytical formula:
num_time_frames = (num_samples_per_clip // hop_length) + 1 = 16000/160 + 1 = 101, which is also what we get from the code.

Data Preprocessing
Label Mapping

We now map all the labels to indices. We first define a set of words that we want to focus on, as wanted_words. All the other labels are mapped to the unknown category.

```
wanted_words = ["yes", "no", "up", "down", "left", "right", "on", "off", "stop", "go"]
unknown_label = "_unknown_"

labels_map = {}
num_classes = 0

def get_label_counts(dataset_instance):
    labels_counter = collections.defaultdict(int)
    for i in range(len(dataset_instance)):
        _, _, label_str, _, _ = dataset_instance[i]
        labels_counter[label_str] += 1
    return labels_counter

all_dataset_labels_counts = get_label_counts(train_data)
all_labels = wanted_words + [unknown_label]

labels_map = {word: index for index, word in enumerate(all_labels)}
num_classes = len(all_labels)
```

Preprocess

We shall now write a function to preprocess each raw_sample, which would comprise the following steps:

1. **Resample**: We first resample the audio to our standard target_sample_rate to ensure all waveforms have a consistent sampling rate.

2. **Pad or Truncate to Target Length**: Each waveform is then padded or truncated to a fixed length, creating equally sized inputs required by the model.

CHAPTER 7 AUDIO CLASSIFICATION WITH PYTORCH

3. **Apply Transforms**: We then apply the Mel spectrogram transform followed by the DB conversion.

4. **Label Mapping**: The label strings are mapped to label indices.

```
def process_raw_sample(raw_waveform, original_sr, label_str, mel_transform, db_transform):
    # Resample if necessary
    processed_waveform = raw_waveform
    if original_sr != target_sample_rate:
      resampler = T.Resample(orig_freq=original_sr, new_freq=target_sample_rate).to(device)
      processed_waveform = resampler(processed_waveform)

    # Pad or truncate to target_len
    if processed_waveform.shape[1] < num_samples_per_clip:
        processed_waveform = torch.nn.functional.pad(processed_waveform,
        (0, num_samples_per_clip - processed_waveform.shape[1]))
    elif processed_waveform.shape[1] > num_samples_per_clip:
        processed_waveform = processed_waveform[:, :num_samples_per_clip]

    # Move to device and apply Mel Spectrogram & DB conversion
    processed_waveform = processed_waveform.to(device)
    mel_spec = mel_transform(processed_waveform)
    mel_spec_db = db_transform(mel_spec)

    # Map label string to integer
    if label_str in labels_map.keys():
      label_idx = labels_map[label_str]
    else:
      label_idx = labels_map[unknown_label]

    return mel_spec_db, label_idx
```

We shall now write a higher-level function that takes the entire dataset (train, validation, or test) and transforms all its samples using the above function.

```python
def create_processed_tensor_dataset(raw_dataset, mel_transform, db_
transform):
    all_mel_db_specs = []
    all_labels = []
    print(f"Processing {len(raw_dataset)} samples...")
    for i in range(len(raw_dataset)): # Limit for faster testing:
    range(min(len(raw_dataset), 500))
        raw_waveform, original_sr, label_str, _, _ = raw_dataset[i]
        spectrogram, label_idx = process_raw_sample(raw_waveform,
        original_sr, label_str, mel_transform, db_transform)
        all_mel_db_specs.append(spectrogram)
        all_labels.append(label_idx)
        if i % 2000 == 0 and i != 0: # Progress update
            print(f"  Processed {i} samples...")
    print("Finished processing samples.")

    stacked_specs = torch.stack(all_mel_db_specs)
    stacked_labels = torch.tensor(all_labels, dtype=torch.long,
    device=stacked_specs.device) # Match device

    return TensorDataset(stacked_specs, stacked_labels)
print("\nProcessing Training Data:")
processed_train_dataset = create_processed_tensor_dataset(train_data, mel_
spectrogram_transform, amplitude_to_db_transform)
print("\nProcessing Validation Data:")
processed_valid_dataset = create_processed_tensor_dataset(val_data, mel_
spectrogram_transform, amplitude_to_db_transform)
print("\nProcessing Test Data:")
processed_test_dataset = create_processed_tensor_dataset(test_data, mel_
spectrogram_transform, amplitude_to_db_transform)
```

Finally, we create three DataLoaders, one for each of the three components of our data (i.e., train, test, and validation).

```
train_loader = DataLoader(processed_train_dataset, batch_size=batch_size,
shuffle=True)
val_loader = DataLoader(processed_valid_dataset, batch_size=batch_size,
shuffle=False)
test_loader = DataLoader(processed_test_dataset, batch_size=batch_size,
shuffle=False)
```

Note that when processing large datasets where the same resampling operation is needed many times, creating the `Resample` object once per unique original sample rate and caching it (e.g., in a dictionary) is a good practice to improve performance. We have not included it in this code for simplicity.

Model Architecture

The model takes as input the image of the Mel spectrogram and aims to classify it into one of the speech labels. Notice that by this formulation, we have reduced the audio command classification problem into an image classification problem! Therefore, we shall use techniques similar to those in Chapter 3; in fact, we shall use the exact same architecture as our final model from Chapter 3.

To recap, the model is composed of three convolution blocks, where each such block contains a 2D convolution layer, followed by a batch norm layer, ReLU activation layer, a dropout layer for regularization, and finally the pooling layer. After these blocks, the volume of feature channels is flattened and passed to a fully connected layer with the ReLU activation.

One of the challenges here is to compute the number of inputs to the fully connected layer after the three convolution blocks. That is achieved by computing the reduced height and width of the input spectrogram due to repeated pooling, where each pooling halves both the width and the height of the image. Hence, we compute the reduced height by the formula: math.floor(math.floor(math.floor(n_mels/2)/2)/2), since n_mels was the original height of the input. We use an analogous formula to compute the reduced width; note that the original width of the input spectrogram is num_time_frames.

```
class SpeechCommandCNN(nn.Module):
    def __init__(self, num_classes, n_mels, num_time_frames):
        super().__init__()
        self.num_classes = num_classes
```

```python
        self.n_mels = n_mels
        self.num_time_frames = num_time_frames

        # First convolution-block
        self.conv1 = nn.Conv2d(
            in_channels=1,
            out_channels=32,
            kernel_size=(3,3),
            padding=1
        )
        self.batch_norm1 = nn.BatchNorm2d(32)
        self.relu_act1 = nn.ReLU()
        self.dropout1 = nn.Dropout(p=0.2)
        self.max_pool1 = nn.MaxPool2d(kernel_size=(2,2))

        # Second convolution-block
        self.conv2 = nn.Conv2d(
            in_channels=32,
            out_channels=64,
            kernel_size=(3,3),
            padding=1
        )
        self.batch_norm2 = nn.BatchNorm2d(64)
        self.relu_act2 = nn.ReLU()
        self.dropout2 = nn.Dropout(p=0.2)
        self.max_pool2 = nn.MaxPool2d(kernel_size=(2,2))

        # Third convolution-block
        final_out_channels = 128
        self.conv3 = nn.Conv2d(
            in_channels=64,
            out_channels=final_out_channels,
            kernel_size=(5,5),
            padding=2
        )
        self.batch_norm3 = nn.BatchNorm2d(128)
        self.relu_act3 = nn.ReLU()
```

```python
        self.dropout3 = nn.Dropout(p=0.2)
        self.max_pool3 = nn.MaxPool2d(kernel_size=(2,2))

        # Flattening layer
        self.flatten = nn.Flatten()

        # Fully connected layers
        # Calculate input features for the fully connected layer
        reduced_height = math.floor(math.floor(math.floor(n_mels/2)/2)/2)
        reduced_width = math.floor(math.floor(math.floor(num_time_frames / 2) / 2) / 2)
        self._fc_input_features = reduced_height*reduced_width*final_out_channels

        self.fc1 = nn.Linear(self._fc_input_features, 128) # Hidden layer
        self.relu_fc1 = nn.ReLU()

        self.fc_output = nn.Linear(128, num_classes) # Output layer (logits)

    def forward(self, x):
        # Input x shape: (batch_size, 1, n_mels, num_time_frames)

        x = self.conv1(x)
        x = self.batch_norm1(x)
        x = self.relu_act1(x)
        x = self.dropout1(x)
        x = self.max_pool1(x)

        x = self.conv2(x)
        x = self.batch_norm2(x)
        x = self.relu_act2(x)
        x = self.dropout2(x)
        x = self.max_pool2(x)

        x = self.conv3(x)
        x = self.batch_norm3(x)
        x = self.relu_act3(x)
        x = self.dropout3(x)
```

```
        x = self.max_pool3(x)

        x = self.flatten(x)

        x = self.fc1(x)
        x = self.relu_fc1(x)

        x = self.fc_output(x) # Output raw logits

        return x
```

Model Training and Evaluation

Model Setup

We first instantiate the model, followed by defining the loss function, optimizer, and learning rate scheduler. Here, we try the ReduceLROnPlateau, which, as the name suggests, reduces the learning rate adaptively as the validation loss plateaus, i.e., stops showing significant improvement for sufficiently many consecutive epochs.

```
# --- Model Instantiation ---
model = SpeechCommandCNN(
    num_classes=num_classes,
    n_mels=n_mels,
    num_time_frames=num_time_frames
).to(device)
print("Model instantiated successfully.")

# --- Loss Function ---
criterion = nn.CrossEntropyLoss()

# --- Optimizer ---
num_epochs = 20
learning_rate = 0.001
optimizer = optim.AdamW(model.parameters(), lr=learning_rate)

# --- Learning Rate Scheduler ---
scheduler = ReduceLROnPlateau(optimizer, mode='min', factor=0.1,
patience=5, verbose=True)
```

Evaluation

We now define our evaluate function to be used during validation and testing. The last argument takes "description" as input, which is meant to contain the purpose for which we are using the function - which, in our case, is either validation or testing.

```python
def evaluate(model, data_loader, criterion, device, description="Validation"):
    model.eval() # Set model to evaluation mode
    running_loss = 0.0
    num_correct_predictions = 0
    total_samples = 0
    start_time = time.time()
    with torch.no_grad():
        for inputs, labels in data_loader:
            inputs, labels = inputs.to(device), labels.to(device)
            outputs = model(inputs)
            loss = criterion(outputs, labels)
            running_loss += loss.item() * labels.size(0)
            predicted = torch.argmax(outputs, dim=1)
            total_samples += labels.size(0)
            num_correct_predictions += (predicted == labels).sum().item()
    end_time = time.time()
    # Handle potential division by zero if data_loader is empty
    if total_samples == 0:
        print(f"Warning: {description} data loader is empty.")
        return 0.0, 0.0

    eval_loss = running_loss / total_samples
    eval_acc = 100.0 * num_correct_predictions / total_samples
    eval_duration = end_time - start_time
    print(f"{description} Loss: {eval_loss:.4f} | {description} Acc: {eval_acc:.2f}% | Duration: {eval_duration:.2f}s")
    return eval_loss, eval_acc
```

Training Loop

We now define the training loop. We use a patience counter for early stopping in case the validation loss shows no improvement for "patience" epochs, similar to Chapter 3. In this case, we also store the model weights that give the lowest validation loss. Once the model finishes training, we load back those stored model weights that gave the best validation accuracy.

```
best_val_accuracy = 0.0
best_model_weights = None # To store the best model state_dict

print("\n-------------------------------------")
print(f"Starting Training for {num_epochs} epochs...")
print("-------------------------------------")
total_start_time = time.time()

patience = 10  # Num epochs to wait for improvement before stopping
patience_counter = 0 # How many epochs have passed without improvement

for epoch in range(num_epochs):
    print(f"\n--- Epoch {epoch+1}/{num_epochs} ---")

    # --- Training ---
    model.train() # Set model to training mode
    running_loss = 0.0
    correct_predictions = 0
    total_samples = 0
    epoch_start_time = time.time()

    for batch_idx, (inputs, labels) in enumerate(train_loader):
        inputs, labels = inputs.to(device), labels.to(device)

        optimizer.zero_grad()
        outputs = model(inputs)
        loss = criterion(outputs, labels)
        loss.backward()
        optimizer.step()

        # Accumulate statistics
        running_loss += loss.item() * inputs.size(0)
```

```python
            predicted = torch.argmax(outputs, dim=1)
            total_samples += labels.size(0)
            correct_predictions += (predicted == labels).sum().item()

    # Handle potential division by zero if train_loader is empty
    if total_samples == 0:
        print("Warning: Training data loader is empty for this epoch.")
        epoch_train_loss, epoch_train_acc = 0.0, 0.0
    else:
        epoch_train_loss = running_loss / total_samples
        epoch_train_acc = 100.0 * correct_predictions / total_samples

    epoch_duration = time.time() - epoch_start_time
    print(f"Epoch {epoch+1} Training Summary:")
    print(f"Training Loss: {epoch_train_loss:.4f} | Training Acc: {epoch_train_acc:.2f}% | Duration: {epoch_duration:.2f}s")

    # --- Validation ---
    val_loss, val_acc = evaluate(model, val_loader, criterion, device, description="Validation")

    scheduler.step(val_loss)

    # --- Early Stopping & Saving Best Model ---
    if val_acc > best_val_accuracy:
        print(f"Validation accuracy improved ({best_val_accuracy:.2f}% -> {val_acc:.2f}%). Saving model weights...")
        best_val_accuracy = val_acc
        best_model_weights = copy.deepcopy(model.state_dict())
        patience_counter = 0 # Reset patience because we found a better model
    else:
        patience_counter += 1

    if patience_counter >= patience:
        print(f"Early stopping triggered after {patience} epochs without improvement on validation accuracy.")
        break
```

```
total_end_time = time.time()
print(f"\n----------- Training Finished -----------")
print(f"Total training time: {(total_end_time - total_start_time)/60:.2f} minutes")
print(f"Best Validation Accuracy achieved: {best_val_accuracy:.2f}%")

print("\nLoading best model weights for final testing...")
model.load_state_dict(best_model_weights)
```

Testing

We again use the evaluate function for testing, passing "Testing" as the description.

```
test_loss, test_acc = evaluate(model, test_loader, criterion, device, description="Testing")
print(f"\nFinal Test Accuracy on holdout data: {test_acc:.2f}%")
```

This gave me a test accuracy of 96.18% when run in the Google Colab environment (using the A100 compute environment).

Once you have reached this stage of implementation, you can congratulate yourself on building an end-to-end system capable of identifying simple human speech commands! You can try to improve the accuracy further by using different approaches, such as including clips with background noise in your dataset (labeling them as "silence" as is common), tuning different hyperparameters, etc.

We will now transition from understanding *what* is said to *how* it is said. Our next project, Speech Emotion Recognition, aims to ignore the content of the spoken words in an audio but tries to isolate the emotions behind the speech. Although this task seems to be orthogonal to what we just worked on, the core methodology – transforming audio into spectrograms and training a neural network to find patterns – is fundamentally the same, showcasing the versatility of the techniques you have learned.

Project 2: Speech Emotion Recognition (SER)

Imagine having digital assistants that understand not only our instructions but also the emotions as we give them. This could revolutionize areas like healthcare, where a digital agent can offer empathy as they understand the patient's concerns, or in interactive learning environments where the AI tutor can adaptively steer the content based on the student's frustration vs. curiosity. This is what we aim to achieve with Speech Emotion

Recognition (SER) – the objective here is to simply gauge the emotion behind the speech while ignoring the sentiments present in the words of the speech themselves. Note that this aims to ignore the words entirely – for example, if a person yells "I am having fun" while sky diving, when he is actually terrified, it should be classified as "afraid" and not so much as "happy".

For training our AI to recognize emotions, we shall use RAVDESS (Ryerson Audio-Visual Database of Emotional Speech and Song), a dataset meticulously crafted by researchers for this very purpose. It features 24 professional actors (12 male, 12 female), uttering the same two sentences, which are fairly neutral in their content:

1. "Kids are talking by the door."

2. "Dogs are sitting by the door."

The professional actors render each of these two sentences in eight different emotions: neutral, calm, happy, sad, angry, fearful, disgust, and surprised. Furthermore, each emotion (except neutral) is expressed at two intensities – normal and strong. Each of these 7*2 + 1 = 15 emotions is rendered twice by each actor. These eight emotions form our labels, which we want to be able to predict, given the emotion-laden spoken utterance.

However, RAVDESS is a relatively small dataset containing only 1440 samples, and using such limited data to predict eight distinct emotions would be challenging. Therefore, to keep things simple, we will group all emotions into three classes – positive (calm, happy, and surprised), negative (sad, angry, fearful, disgust), and neutral (neutral). Our goal will be to predict these broader categories instead.

Note that the neutral wording of these two statements is a deliberate design choice by the researchers to ensure that the AI model learns to discern emotion primarily from prosodic cues – such as pitch, tone, rhythm, and intensity – rather than relying on the sentiment of the words themselves.

Imports

Here are all the imports we would need for this project. We shall see the utility of each of these as we go ahead with the code.

```
# --- Core Python Libraries ---
import os
import glob                  # File handling
```

```python
import math
import random
import time
import copy                    # For saving best model weights
import urllib.request
import zipfile

# --- Numerical & Scientific Computing ---
import numpy as np
import matplotlib.pyplot as plt

# --- PyTorch Core ---
import torch
import torch.nn as nn
import torch.nn.functional as F  # For activation functions
import torch.optim as optim
from torch.optim.lr_scheduler import ReduceLROnPlateau, StepLR

# --- PyTorch Data Utilities ---
from torch.utils.data import Dataset, DataLoader, random_split, Subset, WeightedRandomSampler

# --- Audio Processing (torchaudio) ---
import torchaudio
import torchaudio.transforms as T  # Audio transforms (e.g., MelSpectrogram)

# --- Evaluation and Metrics ---
from sklearn.metrics import confusion_matrix, classification_report
from sklearn.utils.class_weight import compute_class_weight

# --- Interactive Notebook Utilities ---
from IPython.display import Audio, display
```

Dataset

Let us now load the RAVDESS dataset.

```
# Path where RAVDESS will be downloaded and extracted to
DATASET_ROOT_PATH = "./ravdess_data/"

RAVDESS_SPEECH_ZIP_FILENAME = "Audio_Speech_Actors_01-24.zip"
RAVDESS_DOWNLOAD_URL = f"https://zenodo.org/record/1188976/files/{RAVDESS_SPEECH_ZIP_FILENAME}?download=1"
ZIP_FILE_PATH = os.path.join(DATASET_ROOT_PATH, RAVDESS_SPEECH_ZIP_FILENAME)

print(f"Preparing RAVDESS dataset in '{DATASET_ROOT_PATH}' (downloading and extracting)...")

# Ensure the root directory for download exists
os.makedirs(DATASET_ROOT_PATH, exist_ok=True)

# Download the dataset
print(f"Downloading {RAVDESS_SPEECH_ZIP_FILENAME} to {ZIP_FILE_PATH}...")
opener = urllib.request.build_opener()
# Adding a common user-agent can help with some servers
opener.addheaders = [('User-agent', 'Mozilla/5.0')]
urllib.request.install_opener(opener)
urllib.request.urlretrieve(RAVDESS_DOWNLOAD_URL, ZIP_FILE_PATH)

# Extract the dataset
print(f"Extracting {ZIP_FILE_PATH} to {DATASET_ROOT_PATH}...")
with zipfile.ZipFile(ZIP_FILE_PATH, 'r') as zip_ref:
    zip_ref.extractall(DATASET_ROOT_PATH)

print(f"RAVDESS data download and extraction process complete for '{DATASET_ROOT_PATH}'.")
```

The part about "User-Agent" headers may seem cryptic. But it is simply a way of navigating download issues by mimicking requests from a standard web browser.

Custom Dataset Class

In this section, we shall define a custom dataset class RavdessSpeechDataset, which inherits from PyTorch's Dataset class. This requires implementing the methods __init__(), __len__(), and __getitem__().

First, we define some audio processing parameters:

```
# --- Audio Processing Parameters ---
TARGET_SAMPLE_RATE = 16000   # Hz
N_FFT = 400
HOP_LENGTH = 160
N_MELS = 64

CHUNK_DURATION_SECONDS = 4.0
NUM_SAMPLES_PER_CHUNK = int(TARGET_SAMPLE_RATE * CHUNK_DURATION_SECONDS)
NUM_TIME_FRAMES_PER_CHUNK = (NUM_SAMPLES_PER_CHUNK // HOP_LENGTH) + 1
```

Let us see what each of these parameters means:

1. **Target_sample_rate**: The sample rate in Hz for all our audio clips. This means that we shall sample 16000 audio wave amplitudes per second.

2. **N_FFT**: The window size in samples for each FFT, defining the spectrogram's frequency resolution.

3. **HOP_LENGTH**: The number of samples to shift forward between FFT windows, defining time resolution.

4. **N_MELS**: Number of frequency bins in the final MEL spectrogram, defining its height.

5. **CHUNK_DURATION_SECONDS**: The length of each audio clip; longer clips will be truncated, and shorter ones will be padded to this length.

6. **NUM_SAMPLES_PER_CHUNK**: The total number of audio samples in each chunk after resampling and length normalization.

7. **NUM_TIME_FRAMES_PER_CHUNK**: The resulting number of time frames of the Mel spectrogram for each chunk, defining its width.

We will also need to specify the device and a random seed for reproducibility. In case we have cuda, we also set the seed on there as per this random seed.

```
device = torch.device("cuda" if torch.cuda.is_available() else "cpu")

# --- Seed for Reproducibility ---
RANDOM_SEED = 20
torch.manual_seed(RANDOM_SEED)
if torch.cuda.is_available():
    torch.cuda.manual_seed_all(RANDOM_SEED)
```

We now define the label-to-index mapping for each of the eight emotions as per the convention used by this dataset. We also define a mapping from each of the emotions to one of three categories (neutral, positive, negative) as discussed before. These will form our actual target labels for this project.

```
RAVDESS_FILENAME_EMOTION_MAP = {
    1: "neutral", 2: "calm", 3: "happy", 4: "sad",
    5: "angry", 6: "fearful", 7: "disgust", 8: "surprised"
}

# 3-class group mapping
emotion_label_map = {
    "neutral": 0,      # Neutral
    "calm": 1,         # Positive
    "happy": 1,        # Positive
    "surprised": 1,    # Positive
    "sad": 2,          # Negative
    "angry": 2,        # Negative
    "fearful": 2,      # Negative
    "disgust": 2       # Negative
}

index_to_label = {0: "neutral", 1: "positive", 2: "negative"}
num_classes = 3
```

We now define our custom RavdessSpeechDataset class. We rely on the naming convention followed by the dataset here. As you can observe in the downloaded data folder, there are 24 folders, one per actor, labeled from Actor_01 to Actor_24. Each such folder has files labeled with seven numbers. Let us try to understand this with the help of this example filename: 03-01-01-01-01-01-04.wav. Each of the two-digit numbers has the following interpretation in that order:

1. **Modality**: 01 = full-AV, 02 = video-only, 03 = audio-only. Our example file is audio-only as indicated by 03.

2. **Type**: 01 = speech, 02 = song. Our example has speech audio.

3. **Emotion**: 01 = neutral, 02 = calm, 03 = happy, 04 = sad, 05 = angry, 06 = fearful, 07 = disgust, 08 = surprised. Our example has a neutral emotion.

4. **Intensity**: 01 = normal, 02 = strong. Our example is a normal intensity[2].

5. **Statement**: 01 = "Kids are talking by the door", 02 = "Dogs are sitting by the door". Our example contains the first statement.

6. **Repetition**: 01 = first repetition, 02 = second repetition. Our example contains the first repetition.

7. **Actor**: 01 to 24. Odd-numbered actors are male, even-numbered actors are female. Our example is performed by the fourth actor, who is a female.

```
class RavdessSpeechDataset(Dataset):
def __init__(self, base_data_path, target_sr, num_samples_target):
    super().__init__()
    self.base_data_path = base_data_path
    self.target_sr = target_sr
    self.num_samples_target = num_samples_target

    self.file_paths = []
    self.labels = []
    self.actor_ids = []

    # RAVDESS files are originally 48000 Hz
    self.ravdess_original_sr = 48000
```

---

[2] Neutral emotion has only normal intensity.

CHAPTER 7   AUDIO CLASSIFICATION WITH PYTORCH

```python
        self.resampler = None
        if self.ravdess_original_sr != self.target_sr:
            self.resampler = T.Resample(orig_freq=self.ravdess_original_sr,
                                        new_freq=self.target_sr)

        # Walk through actor folders to find all .wav files
        for actor_folder in sorted(os.listdir(self.base_data_path)):
          # E.g. actor_folder can be Actor_02
            actor_path = os.path.join(self.base_data_path, actor_folder)
            if os.path.isdir(actor_path) and actor_folder.
            startswith("Actor_"):
                for file_path in glob.glob(os.path.join(actor_path, "*.wav")):
                    filename = os.path.basename(file_path)
                    # Filename format: 03-01-EMOTION-INTENSITY-STATEMENT-
                    REPETITION-ACTOR.wav
                    parts = filename.split('.')[0].split('-')
                    if len(parts) == 7:
                        vocal_channel = int(parts[1]) # 01 for Speech

                        if vocal_channel == 1: # Filter speech files
                            emotion_id = int(parts[2])
                            # RAVDESS code (1-8)
                            emotion_str = RAVDESS_FILENAME_EMOTION_
                            MAP[emotion_id]
                            label = emotion_label_map[emotion_str]
                            self.labels.append(label)

                            actor_id = int(parts[6]) # Parse actor ID

                            self.file_paths.append(file_path)

                            self.actor_ids.append(actor_id)

        if not self.file_paths:
            print(f"WARNING: No audio files found or processed in {self.
            base_data_path}.")
```

```
    def __len__(self):
        return len(self.file_paths)

    def __getitem__(self, idx):
        file_path = self.file_paths[idx]
        label_index = self.labels[idx]

        waveform, original_sr = torchaudio.load(file_path)

        # Apply the pre-initialized resampler if it exists
        if self.resampler:
            waveform = self.resampler(waveform)

        # Ensure mono channel
        if waveform.shape[0] > 1:
            waveform = torch.mean(waveform, dim=0, keepdim=True)

        # Pad or truncate to target length
        current_len = waveform.shape[1]
        if current_len < self.num_samples_target:
            padding = self.num_samples_target - current_len
            waveform = torch.nn.functional.pad(waveform, (0, padding))
        elif current_len > self.num_samples_target:
            waveform = waveform[:, :self.num_samples_target]

        return waveform, label_index

ravdess_pytorch_dataset = RavdessSpeechDataset(
    base_data_path=DATASET_ROOT_PATH,
    target_sr=TARGET_SAMPLE_RATE,
    num_samples_target=NUM_SAMPLES_PER_CHUNK
)
print(f"\nCustom RavdessSpeechDataset created.")
print(f"Total samples found and processed: {len(ravdess_pytorch_dataset)}")
```

Now, let us print out the waveform shape for the first sample in our dataset.

```
if len(ravdess_pytorch_dataset) > 0:
    example_waveform = ravdess_pytorch_dataset[0]
    print("\nFirst sample from custom Dataset:")
```

```
        print(f"  Waveform shape: {example_waveform.shape}")
        print(f"  Waveform device: {example_waveform.device}")
else:
    print("Warning: Custom RavdessSpeechDataset is empty.")
```

This prints out the following information:

```
First sample from custom Dataset:
Waveform shape: torch.Size([1, 64000])
Waveform device: cpu
```

Defining Transforms

We now define the transforms as in the previous project.

```
mel_spectrogram_transform = T.MelSpectrogram(
    sample_rate=TARGET_SAMPLE_RATE,
    n_fft=N_FFT,
    hop_length=HOP_LENGTH,
    n_mels=N_MELS,
    normalized=False
).to(device)

amplitude_to_db_transform = T.AmplitudeToDB(
    stype="power",  # the input is a power spectrogram
    top_db=80       # Maximum dynamic range in dB
).to(device)
```

Exploratory Data Analysis

To get a better sense of the data, let us now listen to a sample belonging to each of the eight original emotion classes (not just the three classes we use).

```
ORIGINAL_EMOTIONS_LIST = [
    "neutral", "calm", "happy", "sad", "angry",
    "fearful", "disgust", "surprised"
]
```

CHAPTER 7 AUDIO CLASSIFICATION WITH PYTORCH

```python
RAVDESS_EMOTION_MAP = {
    1: "neutral", 2: "calm", 3: "happy", 4: "sad",
    5: "angry", 6: "fearful", 7: "disgust", 8: "surprised"
}
RAVDESS_ORIGINAL_SR = 48000

print(f"--- Finding and Playing one sample for each of the 8 original emotions ---")

# --- 1. Find one file path for each emotion ---
emotion_filepaths = {}
emotions_to_find = set(ORIGINAL_EMOTIONS_LIST)

# Scan through actor folders until we find one example for each emotion
for actor_folder in sorted(os.listdir(DATASET_ROOT_PATH)):
    if not emotions_to_find:
      break # Stop if we've found all of them
    actor_path = os.path.join(DATASET_ROOT_PATH, actor_folder)
    if os.path.isdir(actor_path) and actor_folder.startswith("Actor_"):
        for file_path in glob.glob(os.path.join(actor_path, "*.wav")):
            if not emotions_to_find:
               break

            filename = os.path.basename(file_path)
            emotion_code = int(filename.split('-')[2])
            emotion_str = RAVDESS_EMOTION_MAP.get(emotion_code)

            if emotion_str in emotions_to_find:
                emotion_filepaths[emotion_str] = file_path
                emotions_to_find.remove(emotion_str)

# --- 2. Load, Process, and Play each found audio file ---
print(f"\nPlaying samples (resampled to {TARGET_SAMPLE_RATE} Hz):")

resampler = T.Resample(orig_freq=RAVDESS_ORIGINAL_SR, new_freq=TARGET_SAMPLE_RATE)
```

CHAPTER 7   AUDIO CLASSIFICATION WITH PYTORCH

```
for emotion in ORIGINAL_EMOTIONS_LIST:
    if emotion in emotion_filepaths:
        file_path = emotion_filepaths[emotion]

        # Load and process the waveform
        waveform, sr = torchaudio.load(file_path)
        waveform = resampler(waveform) # Resample
        if waveform.shape[0] > 1: # Ensure mono
            waveform = torch.mean(waveform, dim=0, keepdim=True)

        print(f"\nEmotion: {emotion}")
        display(Audio(data=waveform.numpy(), rate=TARGET_SAMPLE_RATE))
    else:
        print(f"\nCould not find a sample for emotion: {emotion}")
```

This outputs audios in your notebook, one for each emotion. Let us now count the occurrences of each of our labels.

```
labels_list = []
for waveform, label in ravdess_pytorch_dataset:
    labels_list.append(label)

# Count occurrences
counts = np.bincount(labels_list, minlength=len(index_to_label.values()))

print("Class counts per label index:")
for idx, count in enumerate(counts):
    print(f"{idx}: {count} ({index_to_label[idx]})")
```

This prints out the following class counts:

```
Class counts per label index:
0: 96 (neutral)
1: 576 (positive)
2: 768 (negative)
```

This indicates a significant data imbalance, which we shall deal with later.

Train-Test-Validation Split

Usually, we would have split the dataset randomly into train, test, and validation sets in the given proportion. However, we want our ML model to be able to generalize to new people it has never heard of before. To measure this ability honestly, we must prevent the model from "cheating" by just learning to identify specific characteristics of a particular speaker's voice, rather than the general patterns of the emotion itself.

To avoid this, we employ speaker-independent splits, i.e., we ensure that there is no overlap of speakers in the training, validation, and test sets. We achieve this by assigning a group of actors to the training set, a separate group of actors to the validation set, and the remaining group to the test set. This ensures that the model is evaluated fairly on the test set containing actors it has never encountered before. The resulting accuracy might be lower than that of a simple random split, but it is a more realistic and trustworthy measure of how our model would perform in the real world.

```
BATCH_SIZE = 64

all_actor_ids_from_dataset = ravdess_pytorch_dataset.actor_ids
unique_actor_ids = sorted(list(set(all_actor_ids_from_dataset)))

random.seed(RANDOM_SEED) # For reproducible shuffle
random.shuffle(unique_actor_ids)

print(f"Total unique actors found: {len(unique_actor_ids)}")

# 2. Actor assignments for train, validation, and test sets
num_train_actors = 18
num_val_actors = 3
# num_test_actors will be the remainder

train_actors = set(unique_actor_ids[:num_train_actors])
val_actors = set(unique_actor_ids[num_train_actors : num_train_actors + num_val_actors])
test_actors = set(unique_actor_ids[num_train_actors + num_val_actors:])

print(f"Train actors ({len(train_actors)}): {sorted(list(train_actors))}")
print(f"Validation actors ({len(val_actors)}): {sorted(list(val_actors))}")
print(f"Test actors ({len(test_actors)}): {sorted(list(test_actors))}")
```

```
# 3. Indices for each data split based on actor assignments
train_indices = [i for i, actor_id in enumerate(all_actor_ids_from_dataset)
if actor_id in train_actors]
val_indices = [i for i, actor_id in enumerate(all_actor_ids_from_dataset)
if actor_id in val_actors]
test_indices = [i for i, actor_id in enumerate(all_actor_ids_from_dataset)
if actor_id in test_actors]

print(f"\nDataset split sizes (num samples):")
print(f"  Train: {len(train_indices)}")
print(f"  Validation: {len(val_indices)}")
print(f"  Test: {len(test_indices)}")

# 4. Create PyTorch Subset objects
train_dataset_split = Subset(ravdess_pytorch_dataset, train_indices)
val_dataset_split = Subset(ravdess_pytorch_dataset, val_indices)
test_dataset_split = Subset(ravdess_pytorch_dataset, test_indices)
```

During my run, this printed out the following details about the data splits:

```
--- Creating DataLoaders ---
Total unique actors found: 24
Train actors (18): [1, 2, 3, 6, 7, 8, 10, 12, 13, 14, 15, 16, 17, 18, 19,
20, 21, 23]
Validation actors (3): [4, 9, 11]
Test actors (3): [5, 22, 24]

Dataset split sizes (num samples):
  Train: 1080
  Validation: 180
  Test: 180
```

Data Augmentation

To improve our model's generalizability to new speakers and acoustic conditions, we will employ a powerful data augmentation technique called SpecAugment. SpecAugment is applied to the 2D Mel spectrograms just before they are fed into the neural network. It consists of two operations:

1. Frequency masking randomly selects a band of consecutive Mel frequency bins and masks their values, i.e., replaces them with a fixed value (typically the mean value of the entire spectrogram tensor). This would look like a horizontal band getting whitewashed on the 2D Mel spectrogram.

2. Time masking does the same on the time axis, randomly masking out a vertical band on the 2D Mel spectrogram.

By training on these partially masked spectrograms, we force the model to learn more holistic and resilient features from the surrounding context, preventing it from relying excessively on any single, specific acoustic cue.

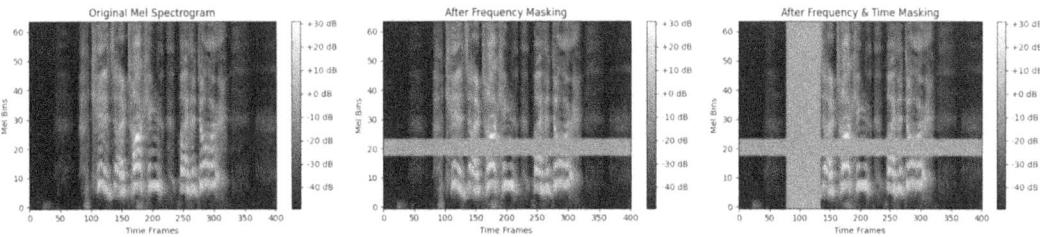

Figure 7-5. *Example of (1) a MEL spectrogram (left), (2) the result after frequency masking (center), and (3) an additional time masking (right)*

```
freq_mask_param = int(N_MELS * 0.05) # Mask 5% Mel bins
time_mask_param = int(NUM_TIME_FRAMES_PER_CHUNK * 0.05) # Mask 5%
time frames

frequency_masking_transform = T.FrequencyMasking(freq_mask_param=freq_mask_
param).to(device)
time_masking_transform = T.TimeMasking(time_mask_param=time_mask_param).
to(device)

print(f"SpecAugment params: Freq mask up to {freq_mask_param} bins, Time
mask up to {time_mask_param} frames.")
```

We now define custom collate functions. We need separate collate functions for training versus evaluation data since we want to apply the SpecAugment method only to the training data. We also normalize the spectrogram using batch-level mean and standard deviation.

```python
# Collate function for training data (applies SpecAugment)
def collate_fn_train(batch):
    waveforms, labels = zip(*batch) # Unzip list of tuples
    waveform_batch = torch.stack(waveforms).to(device)

    spectrogram_batch = mel_spectrogram_transform(waveform_batch)
    spectrogram_batch_db = amplitude_to_db_transform(spectrogram_batch)
    # In your collate_fn (after amplitude_to_db_transform):
    mean = spectrogram_batch_db.mean(dim=[2,3], keepdim=True)
    std = spectrogram_batch_db.std(dim=[2,3], keepdim=True)
    spectrogram_batch_db = (spectrogram_batch_db - mean) / (std + 1e-6)

    # Apply SpecAugment
    spectrogram_batch_db = frequency_masking_transform(spectrogram_batch_db)
    spectrogram_batch_db = time_masking_transform(spectrogram_batch_db)

    label_batch = torch.tensor(list(labels), dtype=torch.long).to(device)
    return spectrogram_batch_db, label_batch

# Collate function for evaluation data (no augmentation)
def collate_fn_eval(batch):
    waveforms, labels = zip(*batch) # Unzip list of tuples
    waveform_batch = torch.stack(waveforms).to(device)

    spectrogram_batch = mel_spectrogram_transform(waveform_batch)
    spectrogram_batch_db = amplitude_to_db_transform(spectrogram_batch)

    # After converting to dB
    mean = spectrogram_batch_db.mean(dim=[2,3], keepdim=True)
    std = spectrogram_batch_db.std(dim=[2,3], keepdim=True)
    spectrogram_batch_db = (spectrogram_batch_db - mean) / (std + 1e-6)

    label_batch = torch.tensor(list(labels), dtype=torch.long).to(device)
    return spectrogram_batch_db, label_batch
```

Handling Data Imbalance

Our data is highly imbalanced, as we saw before during the data analysis stage. We deal with this imbalance by oversampling from the minority class so that the model sees sufficiently many samples from it as well.

We implement this strategy using PyTorch's WeightedRandomSampler. First, we calculate a weight for each emotion class in our training set, with the "balanced" mode automatically assigning higher weights to minority classes. We soften these weights by applying a logarithmic function to them to prevent them from being too extreme. We then assign to each sample the weight corresponding to its class label.

We then pass these weights to an object of type WeightedRandomSampler, which is passed to the DataLoader (which requires setting shuffle=False). During its batchwise iteration, the DataLoader draws samples using this sampler so that samples from minority classes are picked more often. This ensures that each batch of data our model sees during training is approximately balanced across the different emotion categories, forcing the model to learn features for all classes equally.

```
# 1. Get all labels ONLY from your training subset
train_labels = [train_dataset_split.dataset.labels[i] for i in train_
dataset_split.indices]

# 2. Calculate class weights based on the training split's distribution
class_weights_train = compute_class_weight(
    'balanced',
    classes=np.unique(train_labels),
    y=train_labels
)
log_class_weights = np.log1p(class_weights_train)

# 3. Create sample weights for every sample in the training set
sample_weights = torch.tensor([log_class_weights[label] for label in train_
labels])

# 4. Create the WeightedRandomSampler
sampler = WeightedRandomSampler(
    weights=sample_weights,
```

```
        num_samples=len(sample_weights),
        replacement=True
)

# --- Create DataLoaders with appropriate collate_fn ---
train_loader = DataLoader(
    train_dataset_split, batch_size=BATCH_SIZE, shuffle=False,
    collate_fn=collate_fn_train, # Use train collate_fn
    sampler=sampler,
    pin_memory=False
)
val_loader = DataLoader(
    val_dataset_split, batch_size=BATCH_SIZE, shuffle=False,
    collate_fn=collate_fn_eval,
    pin_memory=False
)
test_loader = DataLoader(
    test_dataset_split, batch_size=BATCH_SIZE, shuffle=False,
    collate_fn=collate_fn_eval,
    pin_memory=False
)
```

Model Architecture

Our model's feature extractor comprises four sequential convolution blocks, with each convolution block comprising a convolution layer, followed by batchnorm, a dropout layer, and a maxpool layer.

After these blocks process the input spectrogram, the resulting feature maps must be flattened into a single vector before being passed to our fully connected classifier. To determine the exact size of this vector, we could calculate it analytically using formulas for each layer. However, this manual approach is brittle since the formulas need to be meticulously updated every time the CNN architecture changes.

Instead, we will use a more robust and practical method. We define a helper function, _calculate_cnn_flattened_output_size, that performs a "dry run" by passing a single dummy input tensor through all four convolutional blocks with the same dimensions as our model. It then flattens the resulting feature map and returns the number of features in the final vector (by checking flattened_tensor.shape[1]). This automatically gives us the correct input size for our first fully connected layer, ensuring it always matches our CNN's output.

```
CNN_OUT_CHANNELS = [32, 64, 128, 64]
DROPOUT_RATE = 0.2

class SER_CNN(nn.Module):
    def __init__(self, n_mels, num_time_frames, num_classes,
                 cnn_out_channels, dropout_rate):
        super().__init__()
        # Input dimensions - height: n_mels, width: num_time_frames
        self.n_mels = n_mels
        self.num_time_frames = num_time_frames
        self.num_classes = num_classes

        # Conv block 1
        self.conv1 = nn.Conv2d(1, cnn_out_channels[0], kernel_size=3,
        padding=1)
        self.bn1 = nn.BatchNorm2d(cnn_out_channels[0])
        self.cnn_dropout1 = nn.Dropout2d(p=dropout_rate)
        self.pool1 = nn.MaxPool2d((2, 2))

        # Conv block 2
        self.conv2 = nn.Conv2d(cnn_out_channels[0], cnn_out_channels[1],
        kernel_size=3, padding=1)
        self.bn2 = nn.BatchNorm2d(cnn_out_channels[1])
        self.cnn_dropout2 = nn.Dropout2d(p=dropout_rate)
        self.pool2 = nn.MaxPool2d((2, 2))

        # Conv block 3
        self.conv3 = nn.Conv2d(cnn_out_channels[1], cnn_out_channels[2],
        kernel_size=3, padding=1)
        self.bn3 = nn.BatchNorm2d(cnn_out_channels[2])
        self.cnn_dropout3 = nn.Dropout2d(p=dropout_rate)
        self.pool3 = nn.MaxPool2d((2, 2))
```

CHAPTER 7   AUDIO CLASSIFICATION WITH PYTORCH

```python
        # Conv block 4
        self.conv4 = nn.Conv2d(cnn_out_channels[2], cnn_out_channels[3],
        kernel_size=3, padding=1)
        self.bn4 = nn.BatchNorm2d(cnn_out_channels[3])
        self.cnn_dropout4 = nn.Dropout2d(p=dropout_rate)
        self.pool4 = nn.MaxPool2d((2, 2))

        self.flatten = nn.Flatten()
        self._fc_input_features = self._calculate_cnn_flattened_output_
        size(n_mels, num_time_frames)

        # --- Fully Connected Classifier ---
        self.fc1 = nn.Linear(self._fc_input_features, 128)
        self.relu_fc1 = nn.ReLU()
        self.dropout_fc = nn.Dropout(dropout_rate)
        self.fc_output = nn.Linear(128, num_classes)

    def _calculate_cnn_flattened_output_size(self, n_mels_in, num_time_
    frames_in):
        """ Helper to calculate CNN output flattened size """
        dummy_input = torch.randn(1, 1, n_mels_in, num_time_frames_in)

        # Pass through all conv and pool layers as defined in __init__
        x = self.pool1(F.relu(self.bn1(self.conv1(dummy_input))))
        x = self.pool2(F.relu(self.bn2(self.conv2(x))))
        x = self.pool3(F.relu(self.bn3(self.conv3(x))))
        x = self.pool4(F.relu(self.bn4(self.conv4(x))))
        x_flattened = self.flatten(x)
        return x_flattened.shape[1]

    def forward(self, x):
        # x: (batch, 1, n_mels, num_time_frames)

        x = self.conv1(x)
        x = self.bn1(x)
        x = F.relu(x)
        x = self.cnn_dropout1(x)
        x = self.pool1(x)
```

```
        x = self.conv2(x)
        x = self.bn2(x)
        x = F.relu(x)
        x = self.cnn_dropout2(x)
        x = self.pool2(x)

        x = self.conv3(x)
        x = self.bn3(x)
        x = F.relu(x)
        x = self.cnn_dropout3(x)
        x = self.pool3(x)

        x = self.conv4(x)
        x = self.bn4(x)
        x = F.relu(x)
        x = self.cnn_dropout4(x)
        x = self.pool4(x)

        # Flatten for FC layers
        x = self.flatten(x)

        x = self.fc1(x)
        x = F.relu(x)
        x = self.dropout_fc(x)
        logits = self.fc_output(x)

        return logits
```

Model Training

We first define the relevant training parameters:

```
LEARNING_RATE = 0.0005

# 1. Instantiate the Model
model = SER_CNN(
    n_mels=N_MELS,
    num_time_frames=NUM_TIME_FRAMES_PER_CHUNK,
    num_classes=num_classes,
```

```
    cnn_out_channels=CNN_OUT_CHANNELS,
    dropout_rate=DROPOUT_RATE
).to(device)

print("CNN model instantiated and moved to device.")

criterion = nn.CrossEntropyLoss()
print(f"Loss function: CrossEntropyLoss")

# 3. Define Optimizer
optimizer = optim.AdamW(model.parameters(), lr=LEARNING_RATE)
# ReduceLROnPlateau - reduces LR when a metric has stopped improving.
scheduler = ReduceLROnPlateau(optimizer, mode='max', factor=0.4, patience=3, verbose=True)
print(f"Scheduler: ReduceLROnPlateau (monitoring validation accuracy)")
```

Here, we use the ReduceLROnPlateau scheduler, which, as the name suggests, reduces the learning rate adaptively if the metric being monitored (validation accuracy in our case) doesn't show sufficient improvement for a specified number of "patience" epochs. The intuition is that if we are stuck on a performance plateau, taking smaller, more refined steps might help the optimizer find a better solution.

We now define the evaluate function similar to the previous project, followed by the training loop with early stopping.

```
NUM_EPOCHS = 100

def evaluate(model_to_eval, data_loader, loss_criterion, eval_device,
description="Evaluation"):
    model_to_eval.eval() # Set model to evaluation mode
    running_loss = 0.0
    correct_predictions = 0
    total_samples = 0

    eval_start_time = time.time()
    with torch.no_grad():
        for inputs, labels in data_loader:
            inputs, labels = inputs.to(eval_device), labels.to(eval_
            device).long()
```

```python
            outputs = model_to_eval(inputs)
            loss = loss_criterion(outputs, labels)

            running_loss += loss.item() * inputs.size(0)
            _, predicted = torch.max(outputs.data, 1)
            total_samples += labels.size(0)
            correct_predictions += (predicted == labels).sum().item()

    eval_duration = time.time() - eval_start_time

    eval_loss = running_loss / total_samples
    eval_acc = 100.0 * correct_predictions / total_samples

    return eval_loss, eval_acc, eval_duration

# --- Main Training and Validation Loop ---
best_val_accuracy = 0.0
best_model_weights = copy.deepcopy(model.state_dict())

train_history = {'loss': [], 'accuracy': []}
val_history = {'loss': [], 'accuracy': []}

patience = 10
patience_counter = 0

print(f"\n==================================")
print(f"Starting Training for {NUM_EPOCHS} epochs on {device}...")
print(f"==================================")
total_training_start_time = time.time()

for epoch in range(NUM_EPOCHS):
    print(f"\n--- Epoch {epoch+1}/{NUM_EPOCHS} ---")

    # --- Training Phase ---
    model.train()
    running_train_loss = 0.0
    correct_train_predictions = 0
    total_train_samples = 0
    epoch_train_start_time = time.time()
```

```python
    for batch_idx, (inputs, labels) in enumerate(train_loader):
        inputs, labels = inputs.to(device), labels.to(device).long()

        optimizer.zero_grad()
        outputs = model(inputs)
        loss = criterion(outputs, labels)
        loss.backward()
        optimizer.step()

        running_train_loss += loss.item() * inputs.size(0)
        _, predicted = torch.max(outputs.data, 1)
        total_train_samples += labels.size(0)
        correct_train_predictions += (predicted == labels).sum().item()

    epoch_train_duration = time.time() - epoch_train_start_time
    if total_train_samples > 0:
        epoch_train_loss = running_train_loss / total_train_samples
        epoch_train_acc = 100. * correct_train_predictions / total_
        train_samples
    else:
        epoch_train_loss, epoch_train_acc = 0.0, 0.0

    train_history['loss'].append(epoch_train_loss)
    train_history['accuracy'].append(epoch_train_acc)
    print(f"Training: Loss: {epoch_train_loss:.4f} | Acc: {epoch_train_
    acc:.2f}% | Duration: {epoch_train_duration:.2f}s")

    # --- Validation Phase ---
    print(f"Starting Validation...")
    val_loss, val_acc, val_duration = evaluate(model, val_loader,
    criterion, device, description="Validation")
    val_history['loss'].append(val_loss)
    val_history['accuracy'].append(val_acc)

    # Print validation summary here, using returned values
    print(f"Validation: Loss: {val_loss:.4f} | Acc: {val_acc:.2f}% |
    Duration: {val_duration:.2f}s")
```

```python
    # --- Learning Rate Scheduler Step ---
    scheduler.step(val_acc)

    lr = scheduler.get_last_lr()
    print(f"Learning rate: {lr}")

    # --- Save Best Model & Early Stopping Logic ---
    if val_acc > best_val_accuracy:
        print(f"Validation accuracy improved ({best_val_accuracy:.2f}% -> 
        {val_acc:.2f}%). Saving model state...")
        best_val_accuracy = val_acc
        best_model_weights = copy.deepcopy(model.state_dict())
        patience_counter = 0
    else:
        patience_counter += 1
        print(f"Validation accuracy did not improve for {patience_counter} 
        epoch(s).")

    if patience_counter >= patience:
        print(f"Early stopping triggered after {patience} epochs without 
        improvement.")
        break # Exit the training loop
total_training_end_time = time.time()
print(f"\n----------- Training Finished ({'Early Stopped' if patience_
counter >= patience else 'Completed All Epochs'}) -----------")
print(f"Total training time: {(total_training_end_time - total_training_
start_time)/60:.2f} minutes")
print(f"Best Validation Accuracy achieved: {best_val_accuracy:.2f}%")
```

Note that we save the model parameters for the model with the best validation accuracy so far by saving a deep copy of model.state_dict(). After training, we shall load these model parameters corresponding to the best validation accuracy, instead of using the model from the final epoch.

Model Testing

We shall evaluate the model using our evaluate function used during validation, but this time with the test set. We shall load the best model weights by using the load_state_dict() function.

```
# 1. Load the best model weights (saved from the validation phase during
training)
model.load_state_dict(best_model_weights)

# 2. Evaluate the model on the test set
test_loss, test_acc, test_duration = evaluate(
    model,
    test_loader,
    criterion,
    device,
    description="Final Test"
)

# Print the summary of test results
print(f"\n--- Final Test Results ---")
print(f"Test Loss    : {test_loss:.4f}")
print(f"Test Accuracy: {test_acc:.2f}%")
print(f"Test Duration: {test_duration:.2f}s")
```

This gave me an accuracy of around 60% for my run on the test set.

Confusion Matrix

For many classification tasks, simply evaluating accuracy is not enough, especially when dealing with a high class imbalance. For example, consider the extreme case of credit card fraud detection, where only 1% of transactions are fraudulent. A naive classifier that simply flags all transactions as non-fraudulent would still achieve 99% accuracy while being completely useless as it fails to identify a single fraudulent transaction. Similarly, in medical diagnosis for a rare disease, a model that always predicts "no disease" would be highly accurate but would fail every patient who actually needs help.

In such cases, we need to measure how well our model performs on each class individually. For this, we define metrics called precision, recall, and the F1-score, which are based on the concepts of true positives (TP), false positives (FP), and false negatives (FN).

For defining these concepts, let us think in terms of a class of interest, e.g., the class of "neutral" emotions in our dataset.

1. **True Positive (TP)**: The model predicted neutral, and the label was actually neutral.

2. **False Positive (FP)**: The model incorrectly predicted a positive neutral (i.e., the actual label was not neutral).

3. **False Negative (FN)**: The actual label was neutral, but the model predicted something else.

Based on these, we define prediction, recall, and F1-scores:

1. **Precision**: Precision answers the question: "For all the samples when the model predicted neutral, how many samples were actually neutral?" Mathematically, it is defined as

$$\text{Precision} = \frac{\textit{True Positives}}{\textit{True Positives} + \textit{False Positives}}$$

2. **Recall**: Recall deals with the question: "Of all the samples actually labeled as neutral, how many did the model correctly identify?" It measures the model's ability to find all relevant instances of a class.

$$\text{Recall} = \frac{\textit{True Positives}}{\textit{True Positives} + \textit{False Negatives}}$$

High precision for a class means that when the model identifies that emotion, it is very likely to be correct, with few false alarms, if any. On the other hand, high recall means that the model is good at finding all instances of an emotion. It results in a few missed detections, if any.

Often, there is a trade-off between precision and recall. A model can achieve high recall by being "trigger-happy", i.e., labeling every sample with the class of interest, but this lowers its precision (more false positives). Conversely, a model can be very "cautious" to achieve high precision, but this lowers its recall (more false negatives). The **F1-score** is defined as a way to provide a single metric that balances both precision and recall. It is the **harmonic mean** of the two, i.e.:

$$F1-Score = 2 * \frac{Precision * Recall}{Precision + Recall}$$

The F1-score punishes extreme values more than a simple average. A model must have both high precision and high recall to achieve a high F1-score, making it an excellent metric for evaluating performance on an imbalanced dataset.

Moreover, beyond individual metrics like precision and recall, a **confusion matrix** provides a powerful visual summary of a classification model's performance across all classes. It is a simple matrix where the rows represent the actual true labels and the columns represent the labels predicted by the model. The (i,j)th entry then gives the number of samples that belonged to label i and were classified as label j by the model.

Thus, the numbers along the main diagonal show all the correct predictions, where the predicted label matches the true label. Also, the off-diagonal cells reveal the model's specific weaknesses by showing exactly which classes are being confused for one another – for instance, how many times the model mistook the emotion "positive" for "neutral." Analyzing this matrix is a key step in understanding not just *if* our model is making errors, but *what kind* of errors it's making, which can guide further improvements.

We shall now calculate the precision, recall, and F1-score for each class, along with printing out the confusion matrix. Fortunately, the scikit-learn library makes this straightforward with two key functions: classification_report, which generates a text summary of precision, recall, and F1-score per class, and confusion_matrix, which creates the detailed error matrix itself.

```
# 1. Gather all predictions and true labels from your test set
all_preds = []
ground_truth = []

model.eval()
with torch.no_grad():
    for x, y in test_loader:
        x, y = x.to(device), y.to(device)
        logits = model(x)
```

```
        preds = torch.argmax(logits, dim=1)
        all_preds.extend(preds.cpu().numpy())
        ground_truth.extend(y.cpu().numpy())
all_preds = np.array(all_preds)
ground_truth = np.array(ground_truth)

# 2. Print overall accuracy
accuracy = np.mean(all_preds == ground_truth)
print(f"Test Accuracy: {accuracy:.3f}")

# 3. Confusion Matrix
cm = confusion_matrix(ground_truth, all_preds)
plt.figure(figsize=(3, 3))
plt.imshow(cm, interpolation='nearest', cmap=plt.cm.Blues)
plt.title("Confusion Matrix")
plt.colorbar()
tick_marks = np.arange(num_classes)
plt.xticks(tick_marks, index_to_label.values(), rotation=45)
plt.yticks(tick_marks, index_to_label.values())
plt.xlabel("Predicted Label")
plt.ylabel("True Label")
plt.tight_layout()
plt.show()

# 4. Classification Report (precision, recall, F1)
print("\nClassification Report:")
print(classification_report(ground_truth, all_preds, target_names=index_to_
label.values(), digits=3))
```

Here is what the confusion matrix looked like on my run (see Figure 7-6).

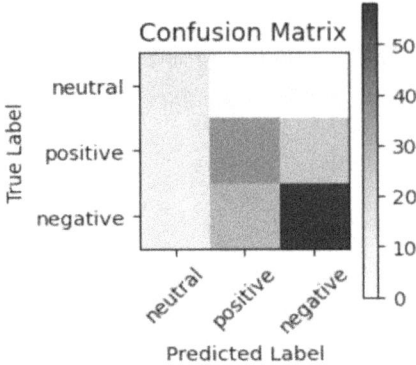

Figure 7-6. *Confusion matrix for our model*

```
Classification Report:
              precision    recall  f1-score   support

     neutral      0.364     1.000     0.533        12
    positive      0.569     0.514     0.540        72
    negative      0.707     0.604     0.652        96

    accuracy                          0.594       180
   macro avg      0.547     0.706     0.575       180
weighted avg      0.629     0.594     0.599       180
```

The classification report contains the precision, recall, and F1-scores for all three classes. Additionally, it also contains the macro average, which is the unweighted average of each metric (precision, recall, and F1-score) for all three classes. The weighted average computes the average weighted by the support of each class, i.e., the number of samples belonging to each class. The accuracy row gives the overall accuracy of the model (it is in the F1-score column purely as a design choice – it has nothing to do with the F1-score directly).

Observe that the neutral class has a recall of 1. This means that the model is trying too hard to ensure that it catches all the neutral examples. Based on these results, it seems that we may have been more aggressive in dealing with the class imbalance than needed. You can try to play around to further scale down these class weights (even more strongly than the log function that we used) and get better results.

Further Improvements

The interested reader can try several advanced techniques to build on this baseline model for Speech Emotion Recognition. Architecturally, the model can be enhanced by converting it into a convolutional recurrent neural network (CRNN) by adding LSTM or GRU layers to model the temporal dynamics of emotional expression. For a more state-of-the-art approach, these recurrent layers could be replaced with a **transformer encoder** to better capture long-range contextual relationships via self-attention.

For more rigorous evaluation, especially with limited data, a **speaker-independent K-fold cross-validation** methodology provides a more reliable estimate of generalization performance. Robustness can also be improved through advanced **data augmentation**, such as mixing in realistic background noise using sounds from an environmental sound dataset like **ESC-50**.

To test the model's generalization on different data, one can experiment with other standard SER datasets like **CREMA-D** and **TESS**. A key strategy for achieving significant performance gains often involves combining these compatible SER datasets (e.g., RAVDESS, CREMA-D, TESS) to create a single, larger, and more diverse training set. This directly addresses data scarcity and significantly improves the model's ability to generalize across different speakers and recording conditions.

Summary

- Sound is a vibration that travels as a pressure wave; two of its core properties are **amplitude** (perceived as loudness) and **frequency** (perceived as pitch).

- To make a continuous sound wave computer-friendly, it is digitized into a **waveform** by **sampling** its amplitude at a regular rate and **quantizing** each sample's value.

- A digitized raw waveform is often transformed into a **Mel spectrogram** – a two-dimensional, image-like representation showing frequency content over time – which is more suitable for ingestion by ML models.

- This transformation effectively reframes audio classification as an **image classification** problem, making it perfect for computer vision techniques such as **convolutional neural networks (CNNs)**.

CHAPTER 7   AUDIO CLASSIFICATION WITH PYTORCH

- In our first project, **speech command recognition**, we use the **Google Speech Commands dataset** to train a model to recognize simple spoken words like "yes," "no," and "up".

- After some primary data preprocessing, we converted each audio wave into a Mel spectrogram image and then used a CNN architecture for classifying the audio wave into one of our speech categories.

- In our second project, we tackled **Speech Emotion Recognition (SER)** on the **RAVDESS dataset**, which contains clips of professional actors speaking the same two sentences with different emotional intonations.

- Similar to our first project, we preprocessed the audio clips and converted them to Mel spectrograms. We then trained a CNN model to recognize the emotions underlying the utterances while ignoring the content of the words themselves.

- A critical concept for the SER project was using **speaker-independent splits** to prevent the model from learning speaker-specific traits and to ensure it generalizes well to new voices.

- To handle the **class imbalance** from grouping emotions, a `WeightedRandomSampler` was used to oversample the minority classes during training.

- **Data augmentation** was used to improve model robustness; specifically, **SpecAugment** was used, which applies random frequency and time masking directly to the spectrograms.

- For imbalanced tasks, accuracy alone can be misleading; we therefore learned about other class-specific metrics such as **precision, recall, F1-score**, and the **confusion matrix. These metrics gave us deeper insights into our model performance.**

- The chapter concluded by highlighting avenues for further improvements, such as evolving the architecture (e.g., to a **CRNN** or **transformer**), using more rigorous evaluation like **K-fold cross-validation**, or **combining multiple datasets**.

CHAPTER 8

Recommender Systems with PyTorch

This chapter introduces the field of recommender systems. It begins by outlining the core objectives of these systems, tracing the historical development of the field, and covering the major approaches employed. The chapter then focuses on the collaborative filtering (CF) paradigm, detailing both user–user CF methods and matrix factorization techniques. Following this background, we delve into a practical project where we work with the MovieLens dataset to predict user movie ratings. It then goes into the coding and accompanying explanations, where we employ the above two CF approaches toward this prediction task.

What Are Recommender Systems

Recommender systems deal with the task of recommending products to users that they are likely to find most interesting. Major web-based consumer companies like Amazon, Netflix, Meta, and Google manage millions of users and huge volumes of data. A key challenge they face is matching each user to the specific items – be it a song or a movie or a product – that they are most likely to engage with. Their main focus remains curating for the users what they want as easily as possible, thereby increasing user engagement. Consequently, recommender systems are integrated into almost every aspect of their **platforms**. Whether it is Netflix recommending movies via its "specially for you" section, or YouTube recommending next videos, or Amazon recommending "You may also like" products, these are all recommender systems at work. They have become ubiquitous in all major big tech companies, and therefore, recommender systems have been an active area of research in ML almost since the inception of the web.

Exploration into recommender systems started in the early 1990s as a way to navigate the information overload created by the growing digital ecosystem. Two approaches became popular initially: content-based filtering, suggesting items similar to a user's known preferences, and collaborative filtering (CF), introduced by systems like Tapestry and GroupLens, which leveraged insights from users with similar tastes. The rise of ecommerce led to even more practical applications, like Amazon using CF to increase discovery and user engagement. A major push to the research in this domain came from the Netflix Prize competition starting in 2006, leading to rapid innovation, particularly popularizing matrix factorization methods. Since then, the field has adopted different types of hybrid approaches, combining different techniques and incorporating context-awareness and side information – often using representation tools like knowledge graphs. More recently, there has been a heavy reliance on deep learning techniques to model complex user-item interactions.

In this chapter, we shall work with two foundational techniques in recommender systems – user–user collaborative filtering and matrix factorization.

Collaborative Filtering

Consider the example of recommending movies. Content-based filtering was the prevalent approach for recommender systems until collaborative filtering came along. This method looked at the *movie's details* – its genre, actors, director, maybe even plot keywords (the item's *metadata*). If you liked action movies starring Matt Damon, it would simply recommend *more* action movies featuring Matt Damon. This approach worked well most times, but it often recommended things very similar to what you already knew.

Collaborative filtering (CF) takes a completely different approach. Instead of relying on the *content* of the items, it focuses on *user behavior* – what did users similar to you actually watch, rate, or buy? The core idea is to leverage the "wisdom of the crowd (similar to you)." It works on the assumption that if many people who liked Movie A also happened to like Movie B, then even if A and B are different genres, there might be some hidden latent connection, and another user who likes A might also enjoy B. It finds patterns purely from a large collection of user ratings or interactions, without needing to know anything about the contents of the movies themselves (like genre, actors, etc.) or about user characteristics (like demographics, etc.).

User–user collaborative filtering is a specific type of CF which, on a high level, finds your "taste twins" – other users who rate items like you do – and suggests things they enjoyed that you haven't seen yet. It is quite intuitive to understand, because it mimics how we get recommendations in real life. Think about asking friends for movie suggestions – you'd probably trust the friend whose movie taste is most similar to yours. User–user CF replicates this:

1. **Your Tastes:** The system first looks at the list of movies *you* have rated and how you rated them.

2. **Find Your Neighbors:** It then searches through the user base to find those who have rated *many of the same movies* in a *very similar way* to you. These users are your "neighbors" or perhaps your "taste twins."

3. **Recommend What Neighbors Liked:** The system then looks at movies that these neighbors rated highly, but which *you haven't rated or seen yet*, and recommends them to you.

The reasoning is simple: if people who have enjoyed the same movies as you also liked this *new* movie, there's a good chance you'll like it too.

Another powerful way to do collaborative filtering is **matrix factorization (MF)**. Instead of directly comparing users to other users, MF takes a different route: it tries to *discover hidden features* – often called **latent factors** – which explain the "reason" why a certain user likes a specific movie. These latent factors are learned about both the users and items, just by considering their interactions. For example, for movies, one factor might represent "seriousness vs. comedy," another "action-packed vs. non-action," etc. For users, there would be corresponding factors that represent how much they prefer each of those movie factors, like whether they prefer "serious vs. comedy" or "action vs. non-action" movies.

The algorithm *learns not only* these latent factors themselves but also which factors it should learn. In our example, the algorithm would figure out that "comedy vs. seriousness" is an important factor to consider while making recommendations and would then learn the inclination of each user and movie to this genre.

Note that the algorithm does not tell us what each latent factor represents – we may have to infer that from the data, looking at which movies are high in a certain factor, and then considering the common characteristics of those movies. In fact, some of the factors may not have any obvious "meaning" or could be a combination of a few characteristics.

Practically, each user and each item is then represented by a vector of floating-point numbers, called an **embedding vector**, where each index of the vector corresponds to a specific learned latent factor.

Once the algorithm learns these embedding vectors, recommending movies to users is surprisingly simple – you simply recommend movies to a user that lie close to it in the latent space. In other words, you calculate the dot-product between the user's embedding vector and the item's embedding vector and use it as a relevance score for recommending the item to the user. So, the more latent factors the user's vector agrees on with the item's vector, the higher will be their dot product and therefore the relevance score. This is ensured during the training process, which looks like this:

1. Start by assigning random embedding vectors of a fixed dimension to all the users and items.

2. Make predictions using the dot product, compare them to the actual known ratings, and calculate the loss function.

3. Use a gradient descent-based optimization algorithm (like Adam) to repeatedly modify the embedding vectors, pushing the predictions gradually toward the actual ratings over many iterations.

After training, these vectors effectively encode the latent factors for users and movies learned based on users' preferences toward different movies.

Recommender Systems Project

We will now delve into our recommender systems project, where we will work with the MovieLens dataset, which contains users' ratings for different movies. Our goal will be to curate good recommendations for our users.

Imports

```
import pandas as pd
import torch
import torch.utils.data as data
import os
```

```
import requests
import zipfile
from io import BytesIO
import time
```

Dataset

We now load the MovieLens dataset. It has different versions available depending on the number of samples in the dataset – here we load the one with 100k samples.

```
class MovieLensDataset(data.Dataset):
    def __init__(self, split="train"):
        super().__init__()
        self.split = split
        self.data_dir = "ml-100k"
        self.data_url = "http://files.grouplens.org/datasets/movielens/" +
        self.data_dir + ".zip"

        ratings_df = self._download_data()

        # Create user and item mappings
        self.user_id_map = {user_id: i for i, user_id in
        enumerate(sorted(ratings_df["user_id"].unique()))}
        self.item_id_map = {item_id: i for i, item_id in
        enumerate(sorted(ratings_df["movie_id"].unique()))}
        self.num_users = len(self.user_id_map)
        self.num_items = len(self.item_id_map)

        # Split data
        train_df, test_df = self._train_test_split(ratings_df)

        if self.split == "train":
            self.data_df = train_df
        elif self.split == "test":
            self.data_df = test_df
        else:
            raise ValueError("Invalid split. Must be 'train' or 'test'.")
```

CHAPTER 8 RECOMMENDER SYSTEMS WITH PYTORCH

```python
        # Convert users, items, ratings in the split to tensors
        self.user_ids = torch.tensor(self.data_df["user_id"].map(self.user_
        id_map).values, dtype=torch.long)
        self.item_ids = torch.tensor(self.data_df["movie_id"].map(self.
        item_id_map).values, dtype=torch.long)
        self.ratings = torch.tensor(self.data_df["rating"].values,
        dtype=torch.float32)

    def _download_data(self):
        if not os.path.exists(self.data_dir):
            print("Downloading MovieLens dataset...")
            response = requests.get(self.data_url, stream=True)
            response.raise_for_status()
            with zipfile.ZipFile(BytesIO(response.content)) as zf:
                zf.extractall()
            print("Download complete.")

        data_file = os.path.join(self.data_dir, "u.data")
        ratings_cols = ["user_id", "movie_id", "rating", "timestamp"]
        ratings_df = pd.read_csv(data_file, sep="\t", names=ratings_cols,
        encoding="latin-1")
        return ratings_df

    def _train_test_split(self, ratings_df, frac=0.8):
        train_df = ratings_df.sample(frac=frac, random_state=28)
        test_df = ratings_df.drop(train_df.index)
        return train_df, test_df

    def __len__(self):
        return len(self.ratings)

    def __getitem__(self, idx):
        return self.user_ids[idx], self.item_ids[idx], self.ratings[idx]
```

Here are some things to note about this code:

1. Since we are extending the Dataset class, we implement our own versions of __init__(), __len__(), and __get_item__() methods as required.

CHAPTER 8 RECOMMENDER SYSTEMS WITH PYTORCH

2. The user ids and movie ids in the dataset may not be contiguous. For our matrix operations later, we would require these to be contiguous, and we therefore create a user_id_map and a movie_id_map to map the ids to contiguous 0-indexed user ids.

3. We use an 80-20 train-test split of the dataset.

Data Analysis and Visualization

We now do a basic analysis of the percentages of ratings in the train and test data.

```
# Get value counts for training set
train_counts = train_dataset.data_df['rating'].value_counts().sort_index()
train_percentages = train_dataset.data_df['rating'].value_
counts(normalize=True).sort_index() * 100

# Get value counts for test set
test_counts = test_dataset.data_df['rating'].value_counts().sort_index()
test_percentages = test_dataset.data_df['rating'].value_
counts(normalize=True).sort_index() * 100

# Print the results
print("\n-------- Training Set --------")
print(f"Total ratings: {len(train_dataset)}")
for rating in train_counts.index:
    print(f"Rating {int(rating)}: {train_counts[rating]:>6} ratings ({train_percentages[rating]:.2f}%)")

print("\n----------- Test Set -----------")
print(f"Total ratings: {len(test_dataset)}")
for rating in test_counts.index:
    # Handle cases where a rating might not exist in the test set (unlikely for 1-5 scale)
    count = test_counts.get(rating, 0)
    percentage = test_percentages.get(rating, 0.0)
    print(f"Rating {int(rating)}: {count:>6} ratings ({percentage:.2f}%)")
print("-" * 30)
```

279

CHAPTER 8   RECOMMENDER SYSTEMS WITH PYTORCH

```python
# --- Visualization ---
# Combine data for easier plotting
plot_df = pd.DataFrame({
    'Rating': train_counts.index,
    'Train Count': train_counts.values,
    'Test Count': test_counts.values,
    'Train %': train_percentages.values,
    'Test %': test_percentages.values
})

fig, ax = plt.subplots(1, 2, figsize=(14, 5))

bar_width = 0.35
index = np.arange(len(plot_df['Rating']))

# Plot Counts of train vs test
rects1 = ax[0].bar(index - bar_width/2, plot_df['Train Count'], bar_width,
label='Train')
rects2 = ax[0].bar(index + bar_width/2, plot_df['Test Count'], bar_width,
label='Test')
ax[0].set_xlabel('Rating')
ax[0].set_ylabel('Number of Ratings')
ax[0].set_title('Rating Counts in Train vs Test Set')
ax[0].set_xticks(index)
ax[0].set_xticklabels(plot_df['Rating'].astype(int))
ax[0].legend()
ax[0].bar_label(rects1, padding=3)
ax[0].bar_label(rects2, padding=3)

# Plot Percentages of train vs test
rects3 = ax[1].bar(index - bar_width/2, plot_df['Train %'], bar_width,
label='Train')
rects4 = ax[1].bar(index + bar_width/2, plot_df['Test %'], bar_width,
label='Test')
ax[1].set_xlabel('Rating')
ax[1].set_ylabel('Percentage of Ratings (%)')
ax[1].set_title('Rating Distribution in Train vs Test Set')
```

```
ax[1].set_xticks(index)
ax[1].set_xticklabels(plot_df['Rating'].astype(int))
ax[1].legend()
# Format percentage labels
ax[1].bar_label(rects3, fmt='%.1f%%', padding=3)
ax[1].bar_label(rects4, fmt='%.1f%%', padding=3)

fig.tight_layout()
plt.suptitle("Rating Frequency Analysis (MovieLens 100k)", y=1.03,
fontsize=16) # Add overall title
plt.show()
```

This gives us the following statistics:

```
-------- Training Set --------
Total ratings: 80000
Rating 1:    4857 ratings (6.07%)
Rating 2:    9094 ratings (11.37%)
Rating 3:   21725 ratings (27.16%)
Rating 4:   27364 ratings (34.21%)
Rating 5:   16960 ratings (21.20%)

----------- Test Set -----------
Total ratings: 20000
Rating 1:    1253 ratings (6.26%)
Rating 2:    2276 ratings (11.38%)
Rating 3:    5420 ratings (27.10%)
Rating 4:    6810 ratings (34.05%)
Rating 5:    4241 ratings (21.20%)
-------------------------------
```

And plots:

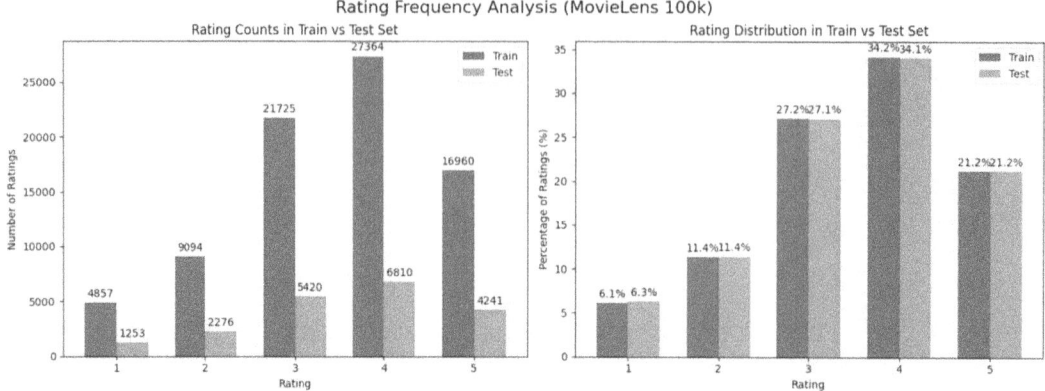

Figure 8-1. *Rating frequencies (left) and percentages (right) in the train (blue) and test (orange) splits of the MovieLens dataset*

As you can see in Figure 8-1, the train and test distributions of ratings are quite similar.

Approach 1: User–User Collaborative Filtering

The first approach we will use is the user–user collaborative filtering, which we described on a high level at the beginning of this chapter. Toward implementing this, we will follow the below plan:

1. Compute the user-item matrix, where rows correspond to users, columns correspond to movies, and each row contains a user's ratings for all the movies.

2. Calculate the user similarity matrix, where the (i,j)th entry contains a score that quantifies the similarity between user i and user j.

3. We shall use the k-Nearest Neighbor (k-NN) algorithm, where, for each user, we find k users (or neighbors) that are closest to it with respect to the above similarity metric.

4. We shall calculate the weighted averages of the rating vectors for these neighbors, weighted by the similarity scores. These weighted averages are then used as predicted movie ratings for movies that the user has not watched yet.

5. There may exist some movies which are not rated by any of the k-nearest neighbors of a user – for such movies, we simply assign the average rating received by that movie as the user rating.

6. Furthermore, there may exist movies that have not been rated by any user yet (typically newly released movies). This is called as the classic "cold-start" problem faced by CF algorithms. For such movies, we simply assign the global average of all the ratings received by all the movies as their rating. This problem only arises initially for each movie, and as more and more users watch that movie, we would be able to assess the "correct" rating of the movie.

7. Another situation we must handle is when new movies (either recently released or new to the platform) enter the system with no ratings. This is typically called the **"cold-start" problem**, when no historical interaction data exists for some items (or users). We deal with this issue by simply assigning the global average rating calculated from all existing user ratings to such movies. As more and more users start rating the movie, the algorithm is able to assess the rating of that movie more accurately.

We shall now go over the details of each of the above steps along with the code.

User Item Matrix

We create a user-item matrix, where the entry (i,j) contains the rating given by user i to movie j. If that movie was not rated by user i, the corresponding matrix entry is 0. Note that the get_item method of our dataset class gets invoked when we index into the dataset as dataset[i]. It returns the triplet of user_idx, item_idx, and the corresponding rating. We then populate this rating into the entry corresponding to user_idx, item_idx.

```
def create_user_item_matrix(dataset):
    """Creates a user-item matrix.
Row i contains ratings given by user i to all the movies
"""
    user_item_matrix = torch.zeros(dataset.num_users, dataset.num_items)
    for i in range(len(dataset)):
```

CHAPTER 8   RECOMMENDER SYSTEMS WITH PYTORCH

```
        user_idx, item_idx, rating = dataset[i]
        user_item_matrix[user_idx, item_idx] = rating
    return user_item_matrix
```

Let us create a small test case with three dummy users and four dummy movie ids to see this function in action.

```
# Define the dimensions for the test case
test_num_users = 3
test_num_items = 4

# Sample rating data: (user_index, item_index, rating)
test_data = [
    (0, 0, 4.0),
    (0, 2, 5.0),
    (1, 0, 3.0),
    (1, 1, 4.0),
    (1, 3, 2.0),
    (2, 1, 1.0),
    (2, 2, 4.0),
    (0, 0, 4.5)
]

# --- Mock Dataset Class ---
class MockDataset:
    def __init__(self, data, num_users, num_items):
        self.data = data
        self.num_users = num_users
        self.num_items = num_items
    def __len__(self): return len(self.data)
    def __getitem__(self, idx): return self.data[idx] # Return tuple directly

# --- Create the Mock Dataset Instance ---
mock_dataset = MockDataset(test_data, test_num_users, test_num_items)

# --- Expected Output Matrix ---
expected_matrix = torch.tensor([
```

```
    [4.5, 0.0, 5.0, 0.0],
    [3.0, 4.0, 0.0, 2.0],
    [0.0, 1.0, 4.0, 0.0]
])
# --- Run the Function with the Mock Dataset ---
print("Running test case...")
actual_matrix = create_user_item_matrix(mock_dataset)

# Assert that the actual matrix is equal to the expected one
assert torch.equal(actual_matrix, expected_matrix), \
        f"Test Failed!\nExpected:\n{expected_matrix}\nActual:\n{actual_matrix}"
# If the assertion passes, print a success message
print("Test Passed: Actual matrix matches expected matrix.")
```

We had to create a mock dataset class to mimic the essential functionalities of our actual dataset class in order to be able to run this test. Ensure that the assert succeeds on your machine, implying that the ratings matrix created matches the expected ratings matrix.

User Similarity Matrix

We use the user-item matrix to obtain the user similarity matrix. The similarity for users u_i and u_j is obtained by taking the cosine similarities of their movie-rating vectors.

We calculate the cosine similarities by first taking the inner product of the user_item_matrix with its transpose. This gives us an n × n matrix, where n is the number of users. Next, we need to divide each entry (i, j) by the norm of user u_i's rating vector and that of user u_j's rating vector. We achieve that by computing the norms of all the user-rating vectors, followed by taking the outer product of these two vectors. Note that the outer product of vectors a and b contain as their (i, j)th entry the product of vector a's ith component and vector b's jth component. Finally, we take an element-wise division by these outer products to get the cosine similarities as intended. I recommend you to work with a small example to ensure that you understand this part of the math.

```
def calculate_user_similarity(user_item_matrix):
    """
    Calculates the user-user similarity.
```

```python
    """
    # 1. Calculate the dot product matrix (numerator)
    # This gives a matrix where element (i, j) is the dot product of user i
    and user j
    dot_product_matrix = torch.matmul(user_item_matrix, user_item_matrix.T)

    # 2. Calculate the norms of each user vector
    user_norms = torch.linalg.norm(user_item_matrix, dim=1)

    # 3. Calculate the outer product of norms (denominator)
    # Gives a matrix s.t. element (i, j) is norm(user_i) * norm(user_j)
# We add 1e-8 to avoid division by zero error
    norm_product_matrix = torch.outer(user_norms, user_norms) + 1e-8

    # 4. Calculate the cosine similarity
    similarity_matrix = dot_product_matrix / norm_product_matrix

    return similarity_matrix
```

We extend our previous dummy test case to ensure the correctness of the calculate_user_similarity function:

```python
# --- Test 2: Verify calculate_user_similarity ---

expected_similarity_matrix = torch.tensor([
    [1.0000, 0.3727, 0.7211],
    [0.3727, 1.0000, 0.1801],
    [0.7211, 0.1801, 1.0000]
])

print("\nRunning Test 2: calculate_user_similarity...")
# Use the *actual* matrix from the first step as input
actual_similarity_matrix = calculate_user_similarity(actual_user_item_matrix)
# Use allclose for floating-point comparisons with a tolerance
assert torch.allclose(actual_similarity_matrix, expected_similarity_matrix, atol=1e-4), \
```

```
        f"Test 2 Failed!\nExpected Similarity Matrix:\n{expected_
similarity_matrix}\nActual:\n{actual_similarity_matrix}"
print("Test 2 Passed.")
```

As before, ensure that the assert succeeds before proceeding any further, to be sure that your calculate_user_similarity function exhibits expected behavior.

Item Averages As Default Values

As discussed in the overview, we need to calculate the average rating received by each item in the case when a movie has not been rated by any of the k-nearest neighbors of a user. For movies with no ratings, we set these item averages to the global average rating, which we calculate in the main function and pass as an argument to the calculate_item_averages function.

```
def calculate_item_averages(user_item_matrix, global_avg_rating):
    """Calculates the average rating for each item."""
    num_items = user_item_matrix.shape[1]
    item_sum_ratings = torch.sum(user_item_matrix, dim=0) # Sum ratings for
    each item
    item_num_ratings = torch.sum(user_item_matrix != 0, dim=0) # Count non-
    zero ratings for each item

    # Initialize averages with global average as fallback
    item_avg_ratings = torch.full((num_items,), global_avg_rating,
    dtype=torch.float32)

    # Create mask for items that have ratings
    has_ratings_mask = item_num_ratings > 0

    # Calculate average only for items which have ratings
    item_avg_ratings[has_ratings_mask] = item_sum_ratings[has_ratings_mask]
/ item_num_ratings[has_ratings_mask]

    return item_avg_ratings
```

Ratings Prediction Using k-NN

We use the k-nearest neighbor algorithm to predict the ratings of a given user for an item. We first use torch's topk function to find the k+1 users most similar to each user. Note that each user will be most similar to itself, and therefore, we need to remove it from the list of top k+1 similar users, trimming the list to k most similar users for each user. We then get the movie ratings given by these k neighbors for the specified item_id (i.e., movie) and return their weighted average. Along the way, we need to apply several filters, like selecting only nonzero ratings (recall that if a user hasn't rated a movie, the corresponding rating value is 0 in our matrix) and using the item average rating computed before if none of the neighbors have rated that movie. We also clamp the rating to fall between 1 and 5, as is required for a valid rating.

```
def predict_rating(user_id, item_id, user_item_matrix, similarity_matrix,
item_avg_ratings, k=20):
    """Predicts the rating for a user and item using k-NN."""

    # 1. Find the k most similar users
    user_similarities = similarity_matrix[user_id]
    topk_similarities, topk_users = torch.topk(user_similarities, k + 1)

    # Remove the user itself from the neighbors
    topk_similarities = topk_similarities[topk_users != user_id][:k]
    topk_users = topk_users[topk_users != user_id] [:k]

    # 2. Get the ratings of the neighbors for the item
    neighbor_ratings = user_item_matrix[topk_users, item_id]

    # 3. Calculate the weighted average
    valid_neighbor_mask = neighbor_ratings != 0  # Boolean mask

    # Apply the mask *only* if there are neighbors who rated the item
    if valid_neighbor_mask.any():
        # Ensure topk_similarities and neighbor_ratings are the same size
        predicted_rating = (topk_similarities[valid_neighbor_mask]
        * neighbor_ratings[valid_neighbor_mask]).sum() / topk_similarities[valid_neighbor_mask].sum()
```

```
        else:
            predicted_rating = item_avg_ratings[item_id] # Use tensor
                directly here
        predicted_rating_clamped = torch.clamp(predicted_rating, 1.0, 5.0)
        return predicted_rating.item()
```

Evaluation

We now write a function to evaluate our algorithm on the test dataset. We calculate the RMSE and MAE losses for the test dataset.

```
def evaluate(test_dataset, user_item_matrix, similarity_matrix, item_avg_
ratings, k=20):
    """Evaluates the model using RMSE and MAE."""
    predictions = []
    actuals = []

    for i in range(len(test_dataset)):
        user_idx, item_idx, actual_rating = test_dataset[i]
        predicted = predict_rating(user_idx, item_idx, user_item_matrix,
            similarity_matrix, item_avg_ratings, k)
        predictions.append(predicted)
        actuals.append(actual_rating.item())

    predictions = torch.tensor(predictions)
    actuals = torch.tensor(actuals)
    rmse = torch.sqrt(torch.mean((predictions - actuals) ** 2))
    mae = torch.mean(torch.abs(predictions - actuals))
    return rmse.item(), mae.item()
```

Putting It All Together

We now have all the functions ready, and it is just a matter of calling them in the right order to see the magic happen.

```
# Create datasets
train_dataset = MovieLensDataset(split="train")
test_dataset = MovieLensDataset(split="test")
```

```python
# Create the user-item matrix *from the training data*
train_user_item_matrix = create_user_item_matrix(train_dataset)

# Calculate user-user similarity
user_similarity_matrix = calculate_user_similarity(train_user_item_matrix)

# Calculate global average (needed for item average fallback)
train_ratings_nonzero = train_user_item_matrix[train_user_item_matrix != 0]
global_average_rating = train_ratings_nonzero.mean().item() if len(train_ratings_nonzero) > 0 else 3.0

# Calculate item averages
item_avg_ratings = calculate_item_averages(train_user_item_matrix, global_average_rating)

# Evaluate
rmse, mae = evaluate(test_dataset, train_user_item_matrix, user_similarity_matrix, item_avg_ratings)
print(f"k-NN (k=20) - Test RMSE: {rmse:.4f}, MAE: {mae:.4f}")
```

This gave me an RMSE of 1.0771 and an MAE of 0.8425 when run on my machine.

Approach 2: Matrix Factorization

We shall now try the matrix factorization approach for the same problem. Here is an overview of the plan that we shall follow:

1. We will start with a matrix factorization class, which contains user embeddings and item embeddings for all our users and items, respectively. We shall initialize them randomly.

2. The forward pass simply computes the dot product of the user and item embeddings of the users and items in that batch.

3. We will then train our model by minimizing the mean squared error (MSE) loss between the predicted and actual ratings.

4. Finally, we shall evaluate our model on the test data as before.

Matrix Factorization Class

We shall now define our matrix factorization class as described before, which contains all the user_embeddings and item_embeddings and simply takes the dot products of the input user_ids and item_ids in the forward method.

```python
class MatrixFactorization(nn.Module):
    def __init__(self, num_users, num_items, embedding_dim=32):
        """
        Initializes the Matrix Factorization model.
        """
        super(MatrixFactorization, self).__init__()

        # --- User and Item Embeddings ---
        self.user_embeddings = nn.Embedding(num_users, embedding_dim)
        self.item_embeddings = nn.Embedding(num_items, embedding_dim)

        # --- Initializing weights ---
        nn.init.xavier_uniform_(self.user_embeddings.weight)
        nn.init.xavier_uniform_(self.item_embeddings.weight)

    def forward(self, user_ids, item_ids):
        """
        Performs the forward pass to predict ratings.
        """
        user_embeds = self.user_embeddings(user_ids)
        item_embeds = self.item_embeddings(item_ids)

        # Calculate dot product (interaction term)
        pred_ratings = (user_embeds * item_embeds).sum(dim=1)

        return pred_ratings
```

Training

Hyperparameters

We first define the hyperparameters. We will be using the Adam optimizer with a weight decay.

```
# --- Hyperparameters ---
EMBEDDING_DIM = 50   # Size of the latent factor vectors
LEARNING_RATE = 0.001 # Learning rate for the optimizer
NUM_EPOCHS = 30

num_users = train_dataset.num_users
num_items = train_dataset.num_items

# --- Instantiate the Model ---
model = MatrixFactorization(num_users, num_items, embedding_dim=EMBEDDING_DIM)

# --- 1.2 Define the Loss Function ---
loss_function = nn.MSELoss()

# --- 1.3 Define the Optimizer ---
weight_decay = 1e-5
optimizer = optim.Adam(
    model.parameters(),
    lr=LEARNING_RATE,
    weight_decay=weight_decay
)
```

Training Loop

We now define the training loop, where we train the model for 30 epochs.

```
print(f"\nStarting training for {NUM_EPOCHS}...")
start_time = time.time()

for epoch in range(NUM_EPOCHS):
    model.train()  # Set the model to training mode
```

```python
    epoch_loss = 0.0
    num_batches = 0

    # Iterate over batches from the training DataLoader
    for user_ids, item_ids, actual_ratings in train_dataloader:
        # 1. Zero out gradients from the previous iteration
        optimizer.zero_grad()

        # 2. Forward pass: Get predictions from the model
        predicted_ratings = model(user_ids, item_ids)

        # 3. Calculate the loss
        loss = loss_function(predicted_ratings, actual_ratings)

        # 4. Backward pass: Compute gradients of the loss w.r.t. model
        parameters
        loss.backward()

        # 5. Optimizer step: Update model parameters based on gradients
        optimizer.step()

        # Accumulate loss for the epoch
        epoch_loss += loss.item()
        num_batches += 1

    # Calculate average loss for the epoch
    avg_epoch_loss = epoch_loss / num_batches

    # Print progress
    print(f"Epoch [{epoch+1}/{NUM_EPOCHS}], Average Loss: {avg_epoch_loss:.4f}")

# --- Training Complete ---
end_time = time.time()
print(f"\nTraining finished in {end_time - start_time:.2f} seconds.")
```

Evaluation

We now write the evaluation loop. After getting the model predictions, we shall clamp them between 1 and 5, as in the first approach.

```
def evaluate(model, dataloader):
    model.eval()  # Set the model to evaluation mode

    all_predictions = []
    all_actuals = []

    # Disable gradient calculations
    with torch.no_grad():
        for user_ids, item_ids, actual_ratings in dataloader:
            # Get predictions
            predicted_ratings = model(user_ids, item_ids)

            # --- Collect predictions and actuals ---
            all_predictions.append(predicted_ratings)
            all_actuals.append(actual_ratings)

    # Concatenate all batch results into single tensors
    all_predictions = torch.cat(all_predictions)
    all_actuals = torch.cat(all_actuals)

    # Clamp predictions to the valid rating range of 1 - 5
    all_predictions = torch.clamp(all_predictions, 1.0, 5.0)

    # Calculate metrics
    rmse = torch.sqrt(torch.mean((all_predictions - all_actuals) ** 2))
    mae = torch.mean(torch.abs(all_predictions - all_actuals))

    return rmse.item(), mae.item() # Return as Python floats
```

We now call the evaluation loop to see the results on the test data.

```
print("Evaluating trained model on the test set...")
test_rmse, test_mae = evaluate_mf(model, test_dataloader)

print(f"\nMatrix Factorization Results:")
print(f"  Test RMSE: {test_rmse:.4f}")
print(f"  Test MAE:  {test_mae:.4f}")
```

On my machine, this gave an RMSE of 1.0792 and MAE of 0.8343, comparable to the first approach of user–user CF.

Note that in our user–user CF approach, we used PyTorch's capabilities purely for vectorized computations. Since there were no learnable parameters, nn.Module, optimizer, etc. were unnecessary. In contrast, matrix factorization **learns** user and item embeddings, so we switch to the full PyTorch training workflow, including nn.Module, embedding, Adam optimizer, etc. This highlights PyTorch's flexibility as both a numerical library and a deep learning framework.

Summary

- Recommender systems aim to curate accurate recommendations for users from a product catalog, which the users are most likely to engage with.

- You can take different approaches toward creating recommendations: **content-based approach**, where recommendations are based on item features (like movie genres, actors, etc.) or user features (like demographics – age, gender, etc.); **collaborative filtering (CF)**, where recommendations are based purely on the historical interactions of users with different items, not taking into consideration any of the product meta data; and other **hybrid approaches**, which combine these and may depend on other techniques like knowledge graphs.

- Collaborative filtering works on the principle that if you agreed with a user in the past on some items, you are likely to agree with them in the future, on new items.

- **User–user collaborative filtering** finds k neighbors with whom you agreed the most on product ratings (or interactions or purchases) and takes the weighted averages of the ratings of these k neighbors to predict your ratings for new items that you haven't interacted with.

- **Matrix factorization** is another collaborative filtering technique, which learns hidden or latent factors for each user and item, such that if a user aligns with an item on sufficiently many latent factors, then it is likely that the user will prefer that item.

- We worked with the MovieLens dataset to predict user ratings using two approaches – user–user CF and matrix factorization.

CHAPTER 9

Image Captioning with PyTorch

In this final chapter, we will embark on what is perhaps our most ambitious project yet, one that sits at the intersection of computer vision and natural language processing. We will combine the skills from our previous work on vision and language models to build an image captioning model, which focuses on generating a relevant caption given any input image.

We will begin with a brief overview of multi-modal models to understand the landscape of AI systems that process multiple data types. From there, we will dive right into our project by preparing the popular Flickr8k dataset. Our journey will cover the complete machine learning pipeline: we will perform exploratory data analysis to get a good sense of our data and then proceed to data preprocessing, using the industry-standard spaCy library for high-quality tokenization and vocabulary building.

The core of our project will be a sophisticated encoder-decoder architecture. We will build an encoder based on a CNN (ResNet-50) model and a decoder based on the classic transformer architecture. Finally, we will train our model and evaluate it using the standard BLEU score metric. We will conclude with a discussion on how this model can be improved even further.

Multi-modal Models

So far, all the models we saw based on an encoder and/or a decoder were text-based models. However, these can also be based on other modalities, such as vision or audio. A system that processes and outputs information from more than one modality is termed a multi-modal model.

There are different types of multi-modal models possible, based on the type of task they perform:

1. **Generation:** Models creating new content in one modality based on input from another.

   - **Image to Text:** For example, the image captioning project we explore in this chapter
   - **Text to Image:** Generating an image given a description
   - **Text-to-Speech:** Generating a speech audio from a given text

2. **Classification and Reasoning:** Different types of reasoning and classification models based on multi-modal inputs/outputs.

   - **Visual Question Answering (VQA):** The model is shown an image and asked a question to answer.
   - **Multi-modal Sentiment Analysis**: The model is given an input video and/or audio and/or text and needs to output a sentiment for it.

3. **Information Retrieval:** These models are designed to find relevant information in one modality using a query from another. They work by learning a shared embedding space where similar concepts are close together.

   - Image Search with Text
   - Video Search with Audio

The Encoder-Decoder Architecture

In Chapter 4, we saw a brief introduction to the encoder-decoder architecture in the context of sequence-to-sequence models, such as for machine translation, summarization, etc. We will now revisit that in the context of multi-modal models.

The fundamental concept remains the same: an encoder network processes the input and compresses it into a compact feature vector, which summarizes the essential "idea" or "thought" present in the input. The decoder network uses this summary and generates a new output sequence.

The crucial difference lies in the *modality* of the input. Previously, our encoder processed a sequence of text; now, it could be required to process an image or an audio. Similarly, the decoder may be required to output a text sequence or an image or an audio, depending on the task.

Any further theoretical exposition would have to be task (or modality) specific. Let us therefore dive right into our image captioning project, learning different concepts about the modalities and the model architecture along the way.

Image Captioning Project

In this project, we aim to train a model that takes an image as its input and outputs a short caption describing the image as accurately as possible. For this, we use the Flickr8k dataset. This dataset contains diverse images, describing one or more people or animals engaged in various activities. For each image, it contains five different human-written captions describing the image.

We will aim to train a model using this data, where it can learn about the characteristics of an image, e.g., who the subjects are (e.g., a boy, or a dog, etc.), what color they are wearing (if humans), and the activity they are engaged in (e.g., running, driving, etc.), along with the descriptions of these characteristics in natural language. Our objective would be that such a trained model should be able to generate a caption describing any new image it sees with sufficient accuracy.

Imports

Here are the imports we will need for this project, bundled together so you don't need to worry about them later:

```
# Standard Python Libraries
import os
import json
import random
import zipfile
from collections import Counter

# Data Handling and Visualization
import pandas as pd
```

Chapter 9 Image Captioning with Pytorch

```
import matplotlib.pyplot as plt
from PIL import Image
from tqdm import tqdm

# NLP and Web
import spacy
import requests

# PyTorch Core
import torch
import torch.nn as nn
import torch.optim as optim
from torch.utils.data import Dataset, DataLoader
from torch.nn.utils.rnn import pad_sequence

# Torchvision
import torchvision.models as models
import torchvision.transforms as transforms
```

Downloading the Data

Let us first download the dataset containing images and their corresponding captions. We thank Jason Brownlee for his Datasets repository: https://github.com/jbrownlee/Datasets, from where we download our data.

```
# --- Configuration ---
image_dataset_url = "https://github.com/jbrownlee/Datasets/releases/download/Flickr8k/Flickr8k_Dataset.zip"
text_dataset_url = "https://github.com/jbrownlee/Datasets/releases/download/Flickr8k/Flickr8k_text.zip"
main_data_folder = "flickr8k_data"

# Create the main destination folder
os.makedirs(main_data_folder, exist_ok=True)

# --- Process the Image Dataset ---
print("Downloading and extracting image dataset...")
image_filename = image_dataset_url.split('/')[-1]
image_zip_path = os.path.join(main_data_folder, image_filename)
```

```
response_img = requests.get(image_dataset_url)
with open(image_zip_path, 'wb') as file:
    file.write(response_img.content)

with zipfile.ZipFile(image_zip_path, 'r') as zip_ref:
    zip_ref.extractall(main_data_folder)

# Remove the image zip file after extraction
os.remove(image_zip_path)
print(f"Removed temporary file: {image_zip_path}")

# --- Process the Text/Annotations Dataset ---
print("Downloading and extracting text dataset...")
text_filename = text_dataset_url.split('/')[-1]
text_zip_path = os.path.join(main_data_folder, text_filename)

response_text = requests.get(text_dataset_url)
with open(text_zip_path, 'wb') as file:
    file.write(response_text.content)

with zipfile.ZipFile(text_zip_path, 'r') as zip_ref:
    zip_ref.extractall(main_data_folder)

# Remove the text zip file after extraction
os.remove(text_zip_path)
print(f"Removed temporary file: {text_zip_path}")
```

The downloaded data resides inside the following structure:

1. Our main data folder is flickr8k_data.

2. It contains a folder called "Flicker8k_Dataset", which contains all the image files.

3. It contains a file "Flickr_8k.trainImages.txt" containing the names of all the images in the training dataset. Similarly, it contains files "Flickr_8k.devImages.txt" and "Flickr_8k.testImages.txt", with image names for the validation and the test dataset, respectively.

4. It contains a file called "Flickr8k.token.txt" containing a mapping of image names to their corresponding captions. Each image has multiple (similar) annotated captions for it, specified with a suffix "#0", "#1", etc. Here are the first ten rows from this file to make this clearer:

```
1000268201_693b08cb0e.jpg#0     A child in a pink dress is climbing up a set of stairs in an entry way .
1000268201_693b08cb0e.jpg#1     A girl going into a wooden building .
1000268201_693b08cb0e.jpg#2     A little girl climbing into a wooden playhouse .
1000268201_693b08cb0e.jpg#3     A little girl climbing the stairs to her playhouse .
1000268201_693b08cb0e.jpg#4     A little girl in a pink dress going into a wooden cabin .
1001773457_577c3a7d70.jpg#0     A black dog and a spotted dog are fighting
1001773457_577c3a7d70.jpg#1     A black dog and a tri-colored dog playing with each other on the road .
1001773457_577c3a7d70.jpg#2     A black dog and a white dog with brown spots are staring at each other in the street .
1001773457_577c3a7d70.jpg#3     Two dogs of different breeds looking at each other on the road .
1001773457_577c3a7d70.jpg#4     Two dogs on pavement moving toward each other .
```

Train-Test-Validation Split

As we just saw, the dataset provides a split into train, test, and validation sets by giving a separate text file containing image names for each split. We read each of these files into a list to help us filter out the data in these three splits later.

```
# Define the base path where the text files are located
text_files_path = "flickr8k_data"
```

```
# Load the list of filenames for each split
with open(os.path.join(text_files_path, "Flickr_8k.trainImages.txt"),
'r') as f:
    train_image_names = [name.strip() for name in f.readlines()]

with open(os.path.join(text_files_path, "Flickr_8k.devImages.txt"),
'r') as f:
    val_image_names = [name.strip() for name in f.readlines()]

with open(os.path.join(text_files_path, "Flickr_8k.testImages.txt"),
'r') as f:
    test_image_names = [name.strip() for name in f.readlines()]

print(f"Training images: {len(train_image_names)}")
print(f"Validation images: {len(val_image_names)}")
print(f"Test images: {len(test_image_names)}")
```

We now have three distinct lists containing the image names for the three data splits.

Reading Captions

We shall now be reading the annotation file, which contains the mapping from each image to multiple captions. We will read it into a pandas dataframe.

```
annotation_file_path = "flickr8k_data/Flickr8k.token.txt"

# 2. Load captions into a pandas DataFrame
captions_df = pd.read_csv(annotation_file_path, sep='\t', header=None,
names=['image_raw', 'caption'])
captions_df['image'] = captions_df['image_raw'].apply(lambda x:
x.split('#')[0])

captions_df.head()
```

CHAPTER 9   IMAGE CAPTIONING WITH PYTORCH

	image_raw	caption
0	1000268201_693b08cb0e.jpg#0	A child in a pink dress is climbing up a set o...
1	1000268201_693b08cb0e.jpg#1	A girl going into a wooden building .
2	1000268201_693b08cb0e.jpg#2	A little girl climbing into a wooden playhouse .
3	1000268201_693b08cb0e.jpg#3	A little girl climbing the stairs to her playh...
4	1000268201_693b08cb0e.jpg#4	A little girl in a pink dress going into a woo...

Figure 9-1. *The image corresponding to the first five captions*

As you can see, the same image, displayed in Figure 9-1, has multiple captions describing it, where the image_raw column contains the suffix "#i" for the ith caption. This dataframe contains image-caption pairs for all the data splits. We will later filter out the rows corresponding to training data or other splits as required.

We create an additional "image" column, which removes the suffix (#0, #1, etc.) and holds just the actual name of each image.

Exploratory Data Analysis (EDA)

Let us print out a few images along with the captions they were annotated with:

```
images_folder = os.path.join(main_data_folder, "Flicker8k_Dataset")

# Group all captions by their corresponding image name
image_to_captions = captions_df.groupby('image')['caption'].apply(list).to_dict()

# Get a list of all unique image names
image_names = list(image_to_captions.keys())

# --- Display Random Samples ---
num_samples_to_show = 3

count = 0
while count < num_samples_to_show:
  random_image_name = random.choice(image_names)

  img_path = os.path.join(images_folder, random_image_name)

  try:
    image = Image.open(img_path)
    plt.figure(figsize=(6, 6))
    plt.imshow(image)
    plt.title(f"Image: {random_image_name}")
    plt.axis('off')
    plt.show()
    count += 1
  except FileNotFoundError:
    print(f"ERROR: Image not found at {img_path}")
    continue

  # --- Display the Captions ---
  print("Captions:")
  captions_for_image = image_to_captions[random_image_name]
  for i, caption in enumerate(captions_for_image):
      print(f"  {i+1}: {caption}")

  print("-" * 80)
```

CHAPTER 9   IMAGE CAPTIONING WITH PYTORCH

Here are the pictures and captions it showed me with this:

```
Image: 218342358_1755a9cce1.jpg
Captions:
  1: A cyclist wearing a red helmet is riding on the pavement .
  2: A girl is riding a bike on the street while wearing a red helmet .
  3: A person on a bike wearing a red helmet , riding down a street .
  4: A woman wears a red helmet and blue shirt as she goes for a bike ride in the shade .
  5: Person in blue shirt and red helmet riding bike down the road
```

```
Image: 1488937076_5baa73fc2a.jpg
Captions:
  1: a few different colored dogs jumping around in the grass behind a house
  2: The Irish setter with the safety vest is running ahead of the Rottwieler and the Dalmation .
  3: Three dogs are running and playing together by the side of a house .
  4: Three dogs running on a lawn .
  5: Three large dogs running quickly next to a house .
```

Image: 3171188674_717eee0183.jpg

Captions:
  1: A city worker with the bright yellow vest is looking intot he crowd of people on the side of a busy street .
  2: A security person in lime green monitors the busy city streets .
  3: A woman wearing a bright yellow jacket and hat stands on a busy street while others walk by .
  4: A woman with a yellow jacket and matching hat watching as people walk by her .
  5: Person in neon green jacket and hat standing on sidewalk while people walk by

Now, let us print out some statistics about the heights and widths of the training images.

```
train_captions_df = captions_df[captions_df['image'].isin(train_image_names)].reset_index(drop=True)

unique_image_names = train_captions_df['image'].unique().tolist()

print(f"Found {len(unique_image_names)} unique images.")

# --- Initialize variables for statistics ---
all_widths = []
all_heights = []
dimension_counts = Counter()

# --- Iterate through all unique images to gather stats ---
for img_name in unique_image_names:
    img_path = os.path.join(images_folder, img_name)
    try:
        with Image.open(img_path) as img:
            width, height = img.size
            all_widths.append(width)
            all_heights.append(height)
            dimension_counts[(width, height)] += 1
    except FileNotFoundError:
```

```
        print(f"Warning: Could not find image {img_name}. Skipping.")
        continue
# --- Calculate and Print Statistics ---
min_width = min(all_widths)
max_width = max(all_widths)
avg_width = sum(all_widths) / len(all_widths)

min_height = min(all_heights)
max_height = max(all_heights)
avg_height = sum(all_heights) / len(all_heights)

# Calculate average aspect ratio (width / height)
aspect_ratios = [w / h for w, h in zip(all_widths, all_heights) if h > 0]
avg_aspect_ratio = sum(aspect_ratios) / len(aspect_ratios)

print("\n" + "="*40)
print("--- Image Dimension Statistics ---")
print("="*40)
print(f"Width:")
print(f"  - Min: {min_width}px")
print(f"  - Max: {max_width}px")
print(f"  - Average: {avg_width:.2f}px")
print("-" * 20)
print(f"Height:")
print(f"  - Min: {min_height}px")
print(f"  - Max: {max_height}px")
print(f"  - Average: {avg_height:.2f}px")
print("-" * 20)
print(f"Average Aspect Ratio (W/H): {avg_aspect_ratio:.2f}")
print("\n--- Top 5 Most Common Dimensions (Width, Height) ---")

for (width, height), count in dimension_counts.most_common(5):
    print(f"  - Size: {width}x{height}, Count: {count}")
print("="*40)

Found 6000 unique images.
```

```
========================================
--- Image Dimension Statistics ---
========================================
Width:
  - Min: 164px
  - Max: 500px
  - Average: 457.88px
--------------------
Height:
  - Min: 127px
  - Max: 500px
  - Average: 397.37px
--------------------
Average Aspect Ratio (W/H): 1.22

--- Top 5 Most Common Dimensions (Width, Height) ---
  - Size: 500x333, Count: 1131
  - Size: 500x375, Count: 965
  - Size: 333x500, Count: 485
  - Size: 375x500, Count: 325
  - Size: 500x334, Count: 244
========================================
```

We also print out the min, max, and average caption lengths in our data.

```
caption_lengths = train_captions_df['caption'].apply(lambda x: len(x.split()))

min_length = caption_lengths.min()
max_length = caption_lengths.max()
avg_length = caption_lengths.mean()

print("--- Caption Length Statistics ---")
print(f"Minimum caption length: {min_length} words")
print(f"Maximum caption length: {max_length} words")
print(f"Average caption length: {avg_length:.2f} words")
```

```
--- Caption Length Statistics ---
Minimum caption length: 2 words
Maximum caption length: 38 words
Average caption length: 11.78 words
```

These statistics will help us make decisions such as the choice of transforms to use on these images, the max length of a caption that needs masking, etc.

Vocabulary and Tokenization

In our chapter on building the language model, we had used a simple Python .split() function for tokenization. It served us well in that case as we were working with a character-level tokenization. With word-level tokenization, however, it fails to correctly handle punctuation and some complex cases. To get a more robust tokenization, we will use **spaCy**, an industry-standard natural language processing (NLP) library. SpaCy's pretrained language models understand English grammar, allowing them to intelligently separate words from punctuation and handle boundary cases.

Introduction to spaCy

Let us quickly go through spaCy's tokenization capabilities with a dummy example.

```
import spacy

# Load the small English model
spacy_model = spacy.load("en_core_web_sm")

# Define a sample sentence
text = "Lizzy's dog isn't running in the park."

# Process the text to create a Doc object
doc = spacy_model(text)

print("\n--- List of Tokens ---")
tokens_list = [token.text for token in doc]
print(tokens_list)

--- List of Tokens ---
['Lizzy', "'s", 'dog', 'is', "n't", 'running', 'in', 'the', 'park', '.']
```

Notice how it handles the same punctuation mark "'" differently: it splits Lizzy's as "Lizzy" and "'s", whereas "isn't" as "is" and "n't". Also, it splits "." into a separate token as we would want.

Vocabulary

We will now build our vocabulary using a spacy model.

```
spacy_model = spacy.load("en_core_web_sm")
word_freq = Counter()
FREQ_THRESHOLD = 5

def tokenizer(text):
    return [token.text.lower() for token in spacy_model.tokenizer(text)]

# Iterate through all captions and count words
for caption in train_captions_df['caption']:
  word_freq.update(tokenizer(caption))

# Create the vocabulary mappings
# Start with special tokens
stoi = {"<PAD>": 0, "<SOS>": 1, "<EOS>": 2, "<UNK>": 3}
idx = 4 # Start indexing for regular words from 4

# Add words that meet the frequency threshold
for word, freq in word_freq.items():
    if freq >= FREQ_THRESHOLD:
        stoi[word] = idx
        idx += 1

# Create the reverse mapping
itos = {index: word for word, index in stoi.items()}

print(f"\nVocabulary size: {len(stoi)}")
```

It prints out that our vocabulary size is 2548.

We create four special tokens:

1. <PAD> for padding

2. <SOS> for "start of sentence"

3. <EOS> for "end of sentence"

4. <UNK> for unknown

For all the other words, we add them to our vocabulary if they occur with frequency exceeding our threshold of 5.

Tokenization and Numericalization

We will now numericalize all our captions text.

```
def numericalize(text):
    tokenized_text = tokenizer(text)
    numericalized = [stoi.get(token, stoi["<UNK>"]) for token in tokenized_text]
    return [stoi["<SOS>"]] + numericalized + [stoi["<EOS>"]]

captions_df['numericalized'] = captions_df['caption'].apply(numericalize)
```

Our captions_df dataframe now contains an additional "numericalized" column, which contains the list of token indices for each caption text. We begin each sequence of token-indices with our special "start-of-sentence" <SOS> token and end it with our "end-of-sentence" token.

Dataset Class

We now define a Dataset class which inherits from the PyTorch Dataset class.

```
class Flickr8kDataset(Dataset):
    def __init__(self, images_folder, captions_df, image_filenames,
    transform):
        self.images_folder = images_folder
        self.transform = transform

        # Filter the df to only include images for this split
        self.df = captions_df[captions_df['image'].isin(image_filenames)]

        # Filter out images with missing files
        self.df['file_exists'] = self.df['image'].apply(lambda img:
        os.path.exists(os.path.join(self.images_folder, img)))
```

```
    self.df = self.df[self.df['file_exists']].reset_index(drop=True)

def __len__(self):
    return len(self.df)

def __getitem__(self, index):
    # Retrieve the image filename and the pre-numericalized caption
    img_id = self.df['image'][index]
    numericalized_caption = self.df['numericalized'][index]

    # Open the image
    img_path = os.path.join(self.images_folder, img_id)
    img = Image.open(img_path).convert("RGB")
    img = self.transform(img)

    return img, torch.tensor(numericalized_caption), img_id
```

The Dataset class is designed to hold our downloaded and preprocessed dataset.

Its **__init__**() method is passed the captions_df dataframe, which contains all the essential information for our dataset. It is also passed an image_filenames list, which contains the list of filenames for images corresponding to one of our data splits (i.e., train, test, or validation). In addition, it is passed the images_folder path so that it can access the actual image files, combining it with an image filename. It is also passed the transforms that need to be performed for each image.

The **__get_item__**() method returns the image, the numericalized caption, and the image id corresponding to the specified index. It converts the raw image into an RGB format first (as expected by our ResNet-50 pretrained model) and applies the specified transforms before returning it.

Image Transforms

```
training_transform = transforms.Compose([
    transforms.RandomResizedCrop(224, scale=(0.8, 1.0), ratio=(0.9, 1.1)),
    transforms.ToTensor(),
    transforms.Normalize(mean=[0.485, 0.456, 0.406],
                         std=[0.229, 0.224, 0.225])
])
```

```
transform = transforms.Compose([
    transforms.Resize(256),
    transforms.CenterCrop(224),
    transforms.ToTensor(),
    transforms.Normalize(mean=[0.485, 0.456, 0.406],
                         std=[0.229, 0.224, 0.225])
])
```

In Chapter 3, we learned about and applied several different transforms to our input images. Here, we apply a few basic of those transforms to our input images. We define two distinct transformation pipelines: one for training and another for validation and testing. For our **training data**, we use RandomResizedCrop, which randomly crops different sections of the image and resizes them to 224×224 in each epoch. This introduces helpful variation in each epoch, encouraging the model to learn more generalizable features. We then normalize it using ImageNet's mean and standard deviation, just like in Chapter 3. Recall that these statistics are widely used in practice since ImageNet contains millions of real-world images and thus provides a representative distribution.

For our **validation and test data**, we use a deterministic pipeline to ensure a consistent evaluation. For this, we use a Resize followed by a CenterCrop. This resizes the smaller edge of the image to 256 pixels and then takes a 224×224 crop from the exact center. Finally, we apply the same normalization using ImageNet statistics.

DataLoaders

As we have seen in several previous projects, the DataLoader is tasked with getting individual samples from the Dataset and assembling them into a batch that can be easily iterated over.

However, the default DataLoader only knows how to handle items that are all the exact same size.

The images in our data are standardized to have the same dimensions, thanks to the transforms we defined. Our image captions, however, have varying lengths. We write a custom collate function to deal with this issue.

```
def collate_fn(batch):
    imgs = [item[0].unsqueeze(0) for item in batch]
    targets = [item[1] for item in batch]
    img_ids = [item[2] for item in batch]
```

```
    imgs = torch.cat(imgs, dim=0)
    targets = pad_sequence(targets, batch_first=False, padding_
value=stoi["<PAD>"])

    return imgs, targets, img_ids
```

Our collate_fn performs two main tasks:

1. **For Images:** It takes the list of individual image tensors of uniform shape [num_channels, width, height], adds a "batch" dimension to each one using .unsqueeze(0), and then stacks them together into a single tensor along dimension 0 using torch.cat(imgs, dim=0). This creates a batch of images of shape [batch_size, num_channels, width, height].

2. **For Captions:** It takes the list of variable-length captions and uses the pad_sequence function to pad the shorter sequences to match the length of the longest caption in the batch.

We now initialize Datasets for each of our splits, and use them to create a DataLoaders for them.

```
# Training Set
train_dataset = Flickr8kDataset(
    images_folder=images_folder,
    captions_df=captions_df,
    image_filenames=train_image_names,
    transform=training_transform,
)
train_loader = DataLoader(
    dataset=train_dataset,
    batch_size=32,
    num_workers=2,
    shuffle=True,  # Shuffle the training data
    pin_memory=True,
    collate_fn=collate_fn
)
```

```python
# Validation Set
val_dataset = Flickr8kDataset(
    images_folder=images_folder,
    captions_df=captions_df,
    image_filenames=val_image_names,
    transform=transform,
)
val_loader = DataLoader(
    dataset=val_dataset,
    batch_size=32,
    num_workers=2,
    shuffle=False, # Do NOT shuffle validation data
    pin_memory=True,
    collate_fn=collate_fn
)

# Test Set
test_dataset = Flickr8kDataset(
    images_folder=images_folder,
    captions_df=captions_df,
    image_filenames=test_image_names,
    transform=transform,
)
test_loader = DataLoader(
    dataset=test_dataset,
    batch_size=32,
    num_workers=2,
    shuffle=False, # Do NOT shuffle test data
    pin_memory=True,
    collate_fn=collate_fn
)
```

```
print(f"Training samples: {len(train_dataset)}")
print(f"Validation samples: {len(val_dataset)}")
print(f"Test samples: {len(test_dataset)}")
Training samples: 30000
Validation samples: 5000
Test samples: 5000
```

In summary, the Dataset provides one sample at a time, comprising an image, the caption, and the image id. The DataLoader manages the process of putting together a batch of samples. For this, it takes help from our custom collate function, which provides the specific instructions on how to intelligently package these samples together.

Model Architecture

We will use the encoder-decoder architecture for this image captioning task (Figure 9-2). The encoder is trusted to "see" the image and create a meaningful representation of the essentials of that image. The decoder is then tasked to read this representation generated by the encoder and generate an appropriate caption, one word at a time, autoregressively.

Encoder Using CNN

The encoder acts as the "eyes" of our system. Given any input image, it is expected to extract essential features of this image and distil them into a numerical vector, called a feature vector. To achieve this, we use transfer learning, i.e., load a powerful pretrained model and fine-tune some of its layers on our data.

CHAPTER 9   IMAGE CAPTIONING WITH PYTORCH

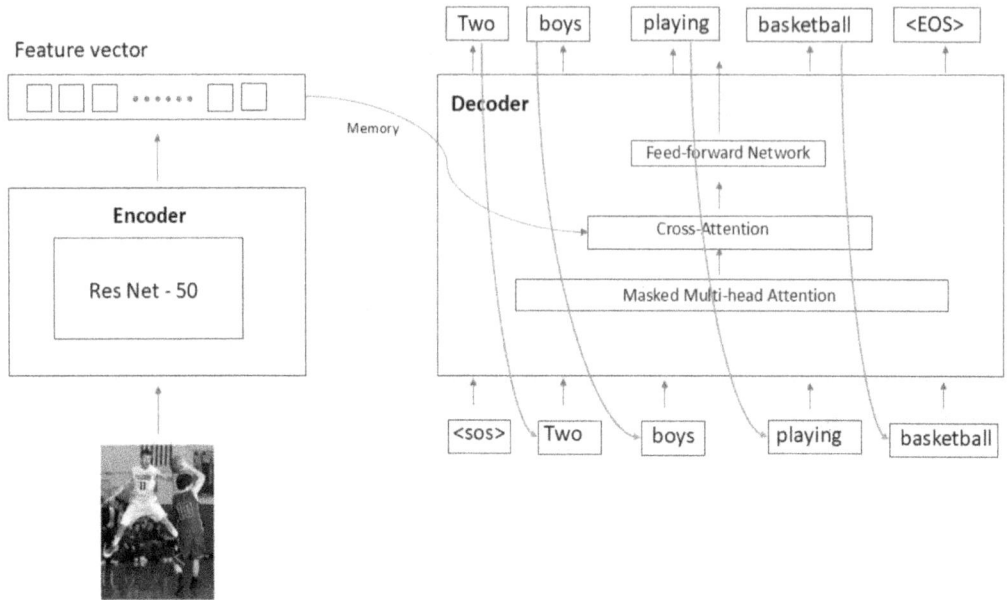

Figure 9-2. A high-level view of the encoder-decoder architecture for our image captioning project

Specifically, we load a pretrained ResNet-50 model, which has been meticulously trained on the ImageNet dataset. We slightly modify it for our specific task by replacing its final classification layer with a new linear layer of our own, which outputs a vector of our desired size, i.e., embed_size. For training, we freeze all the original ResNet layers, ensuring that only our new layer is trained. This process efficiently transforms any input image into a meaningful feature vector of size embed_size.

```
import torchvision.models as models

class EncoderCNN(nn.Module):
    def __init__(self, embed_size):
        super(EncoderCNN, self).__init__()

        # Load the pre-trained ResNet-50 model
        resnet = models.resnet50(weights=models.ResNet50_Weights.DEFAULT)

        for param in resnet.parameters():
            param.requires_grad = False
```

```
        # Replace the final classification layer.
        # NOTE: The parameters of this new layer will have requires_
        grad=True by default.
        resnet.fc = nn.Linear(resnet.fc.in_features, embed_size)

        self.resnet = resnet
        self.relu = nn.ReLU()
        self.dropout = nn.Dropout(0.5)

    def forward(self, images):
        features = self.resnet(images)
        return self.dropout(self.relu(features))
```

In the **__init__** method, we set up all the layers our encoder needs. We first load the ResNet-50 model from the torchvision library, along with its pretrained weights. This model has been trained on the massive ImageNet dataset, making it an expert at recognizing different visual patterns.

We then freeze all the parameters of this pretrained ResNet model, since we don't want to perturb the weights of this expert model.

The pretrained ResNet model has a fully connected classification layer as its final layer, which classifies its input into one of 1000 image classes. We replace this by a fully connected (i.e., linear) layer with dimensions specified by our embed_size parameter. This new layer takes the features from the preceding layer and maps them to a vector of our desired embed_size. The parameters of this new layer will be trainable by default.

Finally, we define a ReLU activation layer and a dropout layer for regularization to follow our ResNet model.

The **forward** method, which defines the data flow, is very straightforward. It simply passes the batch of images through our ResNet network, followed by the ReLU activation and dropout layers.

Decoder Using Transformer

For our task, we use a transformer-based decoder, which acts as the "mouthpiece" of our model. The decoder's main job is to output a relevant caption for the input image using the corresponding feature vectors produced by our EncoderCNN. It produces its output caption one word at a time, and while producing each word, it uses two main sources of information: (1) the feature vector built by the encoder and (2) the words it has output so far.

```python
class TransformerDecoder(nn.Module):
    """
    A Transformer-based decoder
    """
    def __init__(self, embed_size, hidden_size, vocab_size, num_layers, heads):
        super(TransformerDecoder, self).__init__()

        # --- 1. Input and Positional Layers ---
        self.embedding = nn.Embedding(vocab_size, embed_size) # Converts word indices to dense vectors.
        self.positional_encoding = nn.Parameter(torch.zeros(50, 1, embed_size)) # Learnable positional embeddings
        self.dropout = nn.Dropout(0.5)

        # --- 2. The Core Decoder ---
        # Define a single decoder layer
        decoder_layer = nn.TransformerDecoderLayer(
            d_model=embed_size,
            nhead=heads,
            dim_feedforward=hidden_size
        )
        # The main decoder module, which stacks decoder_layers
        self.transformer_decoder = nn.TransformerDecoder(decoder_layer, num_layers=num_layers)

        # --- 3. Output Layer ---
        self.fc_out = nn.Linear(embed_size, vocab_size)

    def forward(self, image_features, captions):
        # --- Step A: Prepare the Caption Input ---
        caption_embeddings = self.embedding(captions)

        # Add positional information to the word embeddings.
        seq_length = captions.shape[0]
        positions = self.positional_encoding[:seq_length]
        x = self.dropout(caption_embeddings + positions)

        # --- Step B: Prepare the Image Input and Masks ---
```

```
# Prepare image features, i.e., the "memory" for the decoder's
cross-attention.
memory = image_features.unsqueeze(0)

# Create a causal mask to prevent the decoder from looking at
future words during training.
tgt_mask = nn.Transformer.generate_square_subsequent_mask(seq_
length).to(image_features.device)

# --- Step C: The Main Decoder Call ---
# Pass all inputs to the decoder module, which handles the layer
looping internally.
predictions = self.transformer_decoder(x, memory, tgt_
mask=tgt_mask)

# --- Step D: Final Output ---
predictions = self.fc_out(predictions)

# Return the raw prediction scores (logits).
return predictions
```

On a high level, the decoder is based upon the attention mechanism, just like in our language modeling project. As it generates each word, it first uses masked self-attention to look at the words it has produced so far, to get a sense of the context and preserve grammatical correctness. It then uses cross-attention to focus on the image features (built by the encoder) most relevant to producing the next word. This allows the model to construct coherent, grammatically correct English sentences that also describe the image accurately enough.

Captions Embedding

The __init__() function begins by defining an **embedding layer** to convert the token indices of all the words in the caption before the current word being generated into an embedding vector. It also has a positional_encoding layer to encode the positional information of each word in the sequence. These positional embeddings are defined using nn.Parameter, ensuring that the model knows to upgrade this layer during the backward pass, making it learnable. It is initialized as a zero-tensor of dimension

50, 1, embed_size, where 50 is an assumed safe upper limit on any caption length, 1 corresponds to the batch dimension (as we will see soon), and embed_size is our specified embedding size.

The forward() function takes as arguments the image features generated by the encoder and the corresponding target caption, tokenized and numericalized. It first generates an embedding of each word (i.e., token) in the target caption. It then slices the positional embedding up to the length of the sequence and adds it to the caption embedding. It then applies a dropout layer on it.

Let us now make sure we understand the math of the dimension gymnastics going on here.

1. The input captions tensor is a 2D tensor of integer indices with the shape (Sequence Length, Batch Size); let's say it is (30, 32).

2. After passing through the embedding layer, the resulting caption_embeddings tensor has shape (Sequence Length, Batch Size, Embed Size), i.e., (30, 32, 256).

3. Our positions tensor, after being sliced, has the shape (Sequence Length, 1, Embed Size), or (30, 1, 256).

4. When we add these two tensors, PyTorch uses a feature called **broadcasting**. It notices that the first and last dimensions match (30 and 256) but the middle dimension does not (32 vs. 1). Broadcasting automatically "stretches" or duplicates the tensor with the dimension of size 1 along that axis to match the larger dimension. In this case, the positional encoding tensor is effectively copied 32 times, allowing the same positional information to be added to every sample in the batch. The final result is a single tensor, x, with the shape (30, 32, 256).

Transformer Decoder Layers

Once the captions have been processed, the model implements a transformer decoder layer using PyTorch's nn.TransformerDecoderLayer (https://docs.pytorch.org/docs/stable/generated/torch.nn.TransformerDecoderLayer.html) class. This is initialized with different parameters of this layer, such as the embedding size, number of heads, and the hidden layer size. These decoder layers are stacked together using PyTorch's nn.TransformerDecoder (https://docs.pytorch.org/docs/stable/generated/torch.nn.TransformerDecoder.html) block. To better understand how these Decoder layers work, observe their usage in the forward() function:

```
predictions = self.transformer_decoder(x, memory, tgt_mask=tgt_mask)
```

It takes as input:

1. **x**: These are the caption embeddings after adding the appropriate positional embeddings, followed by the dropout layer.

2. **memory**: These are the image features constructed by the encoder. We add a dimension to match the format expected by the PyTorch Transformer module.

3. **tgt_mask**: This is a mask implemented using the transformer's generate_square_subsequent_mask method.

The decoder layer is implemented by PyTorch to use these three inputs and internally apply self-attention within x, cross-attention with memory, and then the feed-forward layer, along with layer norms and residual connections, implementing the transformer decoder architecture as specified in the "Attention Is All You Need" paper.

Why Do We Need the Target Mask

During training, we teach the model to generate the entire caption one word at a time. This corresponds to the teacher forcing technique as opposed to the autoregressive technique, where the model is fed the token it itself generates as the next token. A naïve method of achieving this would be to sequentially feed the model one word at a time in a loop.

To speed up this process, we pass it the entire target caption at once and process the prediction of all the words in the caption in parallel using the target mask. For any given word prediction, say, the fourth word, we mask that word and all the words after it. This prevents the model from "peeking" at the fourth word or any subsequent words in the input.

This helps to simulate the real-world scenario where a model has access only to the words it has already predicted (along with the image features), before predicting any given word. In this way, we can train the model on all prefixes of the sentence (<SOS>, <SOS> A, <SOS> A dog, etc.) in parallel, speeding up the training process.

Finally, the transformer outputs are passed through another fully connected layer, which produces a score (a logit) for every single word in our vocabulary. The word with the highest score is the model's prediction for the next word in the sentence.

Image Captioning Encoder-Decoder Model

We combine our EncoderCNN and our TransformerDecoder in this image captioning model.

```
class ImageCaptioningModel(nn.Module):
  """Wrapper model combining the Encoder and Decoder."""
  def __init__(self, embed_size, hidden_size, vocab_size, num_
  layers, heads):
      super(ImageCaptioningModel, self).__init__()
      self.encoder = EncoderCNN(embed_size)
      self.decoder = TransformerDecoder(embed_size, hidden_size, vocab_
      size, num_layers, heads)

  def forward(self, images, captions):
      features = self.encoder(images)
      return self.decoder(features, captions)

  def caption_image(self, image, stoi, itos, max_length=50):
      result_caption = []

      with torch.no_grad():
          x = self.encoder(image)
          captions = torch.tensor([stoi["<SOS>"]]).to(image.device)

          for _ in range(max_length):
              # Prepare the current sequence for the decoder, which expects
              a batch dimension.
              captions_input = captions.unsqueeze(1)

              predictions = self.decoder(x, captions_input)

              # Get the index of the most likely word from the latest
              prediction.
              predicted_word_index = predictions.argmax(2)[-1, :]
```

```
        # Append the new predicted word's index to our sequence for
        the next iteration.
        captions = torch.cat([captions, predicted_word_index], dim=0)

        predicted_word = itos[predicted_word_index.item()]
        result_caption.append(predicted_word)

        # Stop generating if the model predicts the end-of-
        sentence token.
        if predicted_word == "<EOS>":
            break

    return result_caption
```

The __init__() function simply initializes an encoder and a decoder for our model. The forward() function first encodes the input images using our encoder. It then passes these encoded image features along with the target captions to our decoder, which predicts the next word for each prefix of the caption.

The caption_image() function is used for inferencing; it generates a caption given an input image using what is called autoregressive generation. This involves generating the caption one word at a time, where the model bases the prediction of each new word on the image and all the words it predicted before it.

The function begins by passing our single input image through the encoder to get its feature vector, x. It then generates a caption for the image iteratively: first, it passes to the decoder the image encoding x along with just the special <SOS> (start-of-sentence) token, maintained in the `captions` tensor. The decoder predicts the first word of the caption, say "A". The token corresponding to "A" is then appended to the `captions` sequence. Again, we pass to the decoder x along with the updated `captions` sequence comprising tokens corresponding to [<SOS>, "A"]. The decoder then predicts the next word, say "dog", which is appended to `captions` again, and sent back to the decoder. This process is repeated until the decoder outputs the <EOS> (end-of-sentence) token, for a maximum of max_length iterations. The sequence of words predicted in this step-by-step manner is then returned as the predicted caption.

We glossed over one crucial implementation detail in this explanation: the decoder actually outputs logits, i.e., a tensor of scores for all words in the vocabulary. In our implementation, we follow the greedy strategy of selecting the word with the highest logit value as the model's next word prediction.

Let us go over the tensor dimension math here. The function is passed a single image tensor of shape (1, 3, height, width), which our encoder converts into a feature vector x of shape (1, embed_size). Inside the loop, our `captions` tensor, which tracks the generated sequence, starts with a shape of (1) and grows with each new word predicted.

The line captions_input = captions.unsqueeze(1) adds a "batch" dimension, changing the shape from, for example, (4) to (4, 1). This (sequence length, batch size) format is exactly what our decoder expects. The decoder then outputs a `predictions` tensor with the shape (sequence length, 1, vocab size). We use .argmax(2)[-1, :], which works as follows: .argmax(2) finds the index corresponding to the max value along dimension 2. The slicing [-1,:] then selects the last element along dimension 0, corresponding to the prediction using the entire input (i.e., `captions`) sequence. Dimension 1, anyway, has size 1, so selecting all the elements along that dimension selects the single element there. The predicted_word_index is thus a tensor of shape (1). This index is then concatenated back onto our captions tensor, making it one step longer for the next iteration of the loop.

Model Training

```
# --- Configuration ---
device = torch.device("cuda" if torch.cuda.is_available() else "cpu")
print(f"Using device: {device}")
# --- Hyperparameters ---
embed_size = 256
hidden_size = 256
vocab_size = len(stoi)
num_layers = 1
learning_rate = 3e-4
num_epochs = 20
heads = 8

# --- Initialization ---
model = ImageCaptioningModel(embed_size, hidden_size, vocab_size, num_layers, heads).to(device)
criterion = nn.CrossEntropyLoss(ignore_index=stoi["<PAD>"])
optimizer = optim.Adam(model.parameters(), lr=learning_rate)
```

CHAPTER 9 IMAGE CAPTIONING WITH PYTORCH

```python
# --- Training Loop ---
print("\n--- Starting Training ---")
model.train()

for epoch in range(num_epochs):
    print(f"\n--- Epoch [{epoch+1} / {num_epochs}] ---")

    total_epoch_loss = 0.0

    for idx, (imgs, captions, _) in enumerate(train_loader):
        imgs = imgs.to(device)
        captions = captions.to(device)

        # Forward pass
        outputs = model(imgs, captions[:-1, :])
        loss = criterion(outputs.reshape(-1, outputs.shape[2]), captions[1:, :].reshape(-1))

        # Backpropagation
        optimizer.zero_grad()
        loss.backward()
        optimizer.step()

        total_epoch_loss += loss.item()

        # Print progress
        if (idx + 1) % 100 == 0:
            print(f"  Processed Batch {idx + 1} / {len(train_loader)}")

    # Calculate and print the average loss for the epoch
    average_loss = total_epoch_loss / len(train_loader)
    print(f"End of Epoch {epoch+1}, Average Training Loss: {average_loss:.4f}")

print("\n--- Training Complete ---")
```

Model training follows our usual process, similar to all our previous chapters. A noteworthy point is that the loss criterion specifies that the padding token be ignored for the loss function computation. This means that wherever the target caption contains a <PAD> token, that position is completely excluded from the loss computation, irrespective of the model's prediction for that position. Note, however, that if the target contains a non-pad token at any position, but the model predicts a <PAD>, the model is penalized for such a prediction. This ensures that the model focuses entirely on predicting the actual words in the caption and isn't rewarded for its ability to predict the artificial padding tokens we added for batching.

Model Inference: Generating Captions for Images

```
model.eval()
def generate_caption(image_path, model, transform, stoi, itos, device):
    """Loads an image, preprocesses it, and generates a caption."""
    image = Image.open(image_path).convert("RGB")
    image_tensor = transform(image).unsqueeze(0).to(device)

    # The model's caption_image method generates the sequence of indices
    caption_indices = model.caption_image(image_tensor, stoi, itos)

    # Convert indices to words, filtering out special tokens
    caption_words = [word for word in caption_indices if word not in ["<SOS>", "<EOS>", "<PAD>"]]
    return ' '.join(caption_words)

# --- Select a few random test images to display ---
num_samples_to_show = 3
random_test_images = random.sample(test_image_names, num_samples_to_show)
images_folder = "flickr8k_data/Flicker8k_Dataset"

print(f"\n--- Generating Captions for {num_samples_to_show} Random Test Images ---")

for img_name in random_test_images:
    img_path = os.path.join(images_folder, img_name)

    # Generate the caption
    generated_caption = generate_caption(img_path, model, transform, stoi, itos, device)
```

```
# Display the results
image = Image.open(img_path)
plt.figure(figsize=(7, 7))
plt.imshow(image)
plt.title(f"Generated Caption:\n{generated_caption}", wrap=True,
fontsize=12)
plt.axis('off')
plt.show()
print("-" * 80)
```

We demonstrate inferencing on three randomly chosen images in the test set. We write a generate_caption function to generate captions for any given image. Internally, it calls the model's caption_image method to generate the model's predicted caption.

Here are the three images it picked on my run, along with the captions that the model generated.

CHAPTER 9 IMAGE CAPTIONING WITH PYTORCH

Generated Caption:
a white crane is landing in the water .

Generated Caption:
a man in a red shirt is climbing a rock .

As you can see, it generated fairly accurate captions for the first two images. It performed poorly on the third image, though, not even getting the shirt color right.

However, note that even for the third image, it did generate a coherent sentence, even though it was not suitable for the image. This is commendable because, while our encoder CNN was pretrained, we trained our decoder transformer from scratch. It learned to generate these coherent captions just from our training process!

Model Evaluation

The BLEU Score

To evaluate our model's performance on the held-out test data, we use a metric called the **BLEU (Bilingual Evaluation Understudy) score**[1]. The BLEU score was originally designed for evaluating the machine translation task. It works by comparing a model-generated caption to a set of human-written reference captions.

It first calculates the **unigram precision**, i.e., the fraction of unigrams (i.e., single words) from the model-generated caption that also belonged to one of the reference captions. For example:

Model generated: The cat sat on the the the mat.

Reference: The dog sat on the mat.

Here, seven out of the eight words in the model-generated caption belong to the reference, so the unigram precision would be 7/8.

However, it is clear to us that the model should be penalized for the extra "the"s that it generated. For this, the modified precision is defined, which "consumes" any word in the reference that has been matched to some word in our model-generated sequence. In this case, the two "the"s in the reference will be consumed first, and two of the four "the"s in the model-generated sentence would remain unmatched. The modified unigram precision of this would therefore be only 5/8.

In an analogous way, we define modified bigram (two-gram), trigram (three-gram), and quadrigram (four-gram) precisions. One component of the BLEU score is then computed using a geometric mean of these four precisions.

One caveat with this would be that if the model generated a short sentence like "the mat", then since both of these words can be found in the reference, its precision would all be 1, and the BLEU score would be 1 as well, which is the best possible score! This issue

---

[1] The BLEU score was first introduced in this paper: https://aclanthology.org/P02-1040.pdf.

is handled by penalizing very short outputs by defining a multiplicative **brevity penalty** (BP). Any input shorter than r, i.e., the length of the reference sentence that is closest to the length of the generated sentence, is penalized as

$$BP = 1 \quad \text{if } c > r$$

$$= e^{\left(1 - \frac{r}{c}\right)} \quad \text{if } c \leq r$$

The final BLEU score is computed by multiplying this BP by the geometric mean of the modified precisions of the four n-grams (unigram, two-gram, three-gram, four-gram).

Smoothing Function

A strict n-gram comparison can be harsh, especially for shorter sentences where a four-gram match might not occur, leading to a score of zero. To handle this, we use a SmoothingFunction proposed by Chen and Cherry. This function applies a small mathematical adjustment to avoid these zero scores, providing a more stable evaluation across all captions.

Model Testing

We now write code to evaluate our model using this BLEU score.

```
from nltk.translate.bleu_score import sentence_bleu, SmoothingFunction

print("\n--- Evaluating Model on Test Set ---")

model.eval()
all_bleu_scores = []
chencherry = SmoothingFunction()

with torch.no_grad():
    for imgs, captions, img_ids in test_loader:
        imgs = imgs.to(device)

        # Process each image in the batch
        for i in range(len(imgs)):
```

```
            single_img = imgs[i].unsqueeze(0) # Get one image from
            the batch
            current_img_id = img_ids[i]

            # Generate a caption for the single image
            generated_caption_tokens = model.caption_image(single_img,
            stoi, itos)

            # Get the 5 human-written reference captions
            reference_captions_text = image_to_captions[current_img_id]

            reference_tokens = [tokenizer(caption) for caption in
            reference_captions_text]

            # Calculate the BLEU-4 score
            bleu_4_score = sentence_bleu(
                reference_tokens,
                generated_caption_tokens,
                weights=(0.25, 0.25, 0.25, 0.25),
                smoothing_function=chencherry.method1
            )

            all_bleu_scores.append(bleu_4_score)

# Calculate the average BLEU score across the entire test set
average_bleu_score = sum(all_bleu_scores) / len(all_bleu_scores)

print("\n" + "="*50)
print(f"Total Test Images Evaluated: {len(all_bleu_scores)}")
print(f"Average BLEU-4 Score on the Test Set: {average_bleu_score:.4f}")
print("="*50)

==================================================
Total Test Images Evaluated: 5000
Average BLEU-4 Score on the Test Set: 0.1509
==================================================
```

The evaluation code works by looking at one image at a time. For each image, we first use our trained model to generate a single caption. Then, we look up the five different human-written reference captions for that image. We then call the NLTK library's sentence_bleu function, passing our model-generated caption, along with a tokenized version of these five reference captions. We pass a weights vector, which signifies the weights to be used while taking the geometric mean of unigram, bigram, trigram, and quadrigram precision. We keep all these weights equal (0.25 each), signifying that all four n-grams are given equal importance while computing the BLEU score. Finally, we take the average BLEU score for all the images in our test data.

For my run, after training for 20 epochs, it gave a BLEU score of about 0.15, which is a decent score given the resources we used for this project.

Further Improvements

Here are some simple yet effective ways to improve the performance and quality of our image captioning model:

1. **Use a Stronger Encoder**: Replace ResNet-50 with a more advanced model like EfficientNet, Vision Transformers (ViTs), or CLIP for richer visual features.

2. **Fine-Tune the Encoder**: Instead of freezing all the CNN layers, allow some or all to be trained, possibly with a lower learning rate.

3. **Beam Search Decoding**: During inference, replace greedy decoding with beam search to generate more fluent and accurate captions. Beam search keeps track of a "beam" of, say, ten best hypotheses at each step, producing more fluent and accurate captions.

4. **Tune Hyperparameters**: Try changing the different model hyperparameters, such as embedding size, number of decoder layers, or attention heads; experiment with learning rate schedules.

5. **Pretrained Word Embeddings**: Initialize the decoder's embedding layer with GloVe or FastText to leverage prior linguistic knowledge.

6. **Scheduled Sampling**: Gradually let the decoder use its own predictions during training to reduce reliance on teacher forcing.

Each of these changes can be implemented with minimal additional code and can lead to noticeable improvements in BLEU scores and caption quality.

Summary

- Multi-modal models are AI systems designed to process and relate information from multiple different data types, such as text and images (or audio), simultaneously.

- The encoder-decoder architecture is a popular technique used in many multi-modal models, where the encoder distils the "thought" in the input – whatever the modality – and the decoder uses this thought to generate a suitable output in the required modality.

- We embarked on an image captioning project using the popular Flickr8k dataset, which contains images paired with five descriptive human-written captions each, describing them.

- For data preprocessing, we used spaCy, an industry-standard NLP library that provides robust tokenization to correctly separate words and punctuation.

- Our data pipeline comprised a custom PyTorch `Dataset` class to load image-caption pairs, a custom `collate_fn` to handle padding of variable-length captions, and the `DataLoader` to efficiently create batches for training.

- We used standard image transforms from the `torchvision` library to resize, crop, and normalize the image data, ensuring it matched the input requirements of our pretrained model.

- Our model was based on the popular encoder-decoder architecture, where the encoder was responsible for encapsulating the essence of the input image into a feature vector, whereas the decoder uses this feature vector to generate an appropriate caption for it.

- For the encoder, we used a pretrained ResNet-50 model as a feature extractor to understand the visual content of the images, replacing its final classification layer to output a vector of our specified size. We froze all the layers of the pretrained model, except this last one, during training.

- For the decoder, we implemented a transformer architecture, which leverages masked self-attention and cross-attention mechanisms, along with feed-forward layers.

- The decoder generates a suitable caption one word at a time. During training, it uses teacher-forcing, i.e., it tries to predict each word of the target caption simultaneously in parallel while using a mask to ensure that it is only allowed to look at the words that came before it. During inference, it uses autoregression, where it is fed its own previously predicted words in order to generate the next word in the sequence.

- We evaluated our model on a held-out test set using the standard BLEU score metric, which compares the machine-generated captions to the human-written references using a combination of precision on unigrams, bigrams, trigrams, and quadrigrams.

- We concluded by discussing several avenues for future improvement, such as using more advanced encoders, implementing beam search decoding, and tuning hyperparameters to enhance caption quality.

Index

A

Activation function, 5–8, 55, 72
AGNews dataset, 133
AGNews training data, 110
AlexNet, 53
Amplitude
    frequency, 220
    higher amplitude, 220
    lower amplitude, 220
AmplitudeToDB transform, 228
Artificial intelligence (AI), 1–2, 21, 242
Artificial ordering, 96
Attention-based model, 197
Autograd mechanism, 23, 36, 38, 49
Automated text summarization
    abstractive, 154, 158
    conversational chat summarizer
        compute metrics function, 165–167
        dataset, 159
        EDA, 159–162
        Google's T5 model, 158
        imports, 158, 159
        inference, 170–172
        model testing, 170
        model training, 167–170
        preprocess, 162, 163
        results, 171
        ROUGE, 164
        tokenization, 164
    extractive, 153, 154
    Wikipedia article, 154

API, 154
BART model, 156
chunking, large documents, 156, 157
Hugging Face pretrained model, 155
summarizer, 155
summary text, 155
Autoregressive generation, 193, 197, 325
Autoregressive process, 217
AutoTokenizer, 128, 147

B

Batching, 106, 148, 190, 191
Batch normalization layer, 59–60, 74, 78, 91
Beam search, 170, 334
Bidirectional Encoder Representations from Transformers (BERT) model, 103, 128–132
Bigram model, 177, 178
    architecture, 192, 193
    evaluate function, 197–199
    generate text, 193–197
    testing, 201
    text generation, 202, 203
    training, 199–201
Bilingual Evaluation Understudy (BLEU) score, 331–332
Broadcasting, 322

INDEX

C

_calculate_cnn_flattened_output_size, 258
calculate_item_averages function, 287
calculate_user_similarity function, 286, 287
caption_image() function, 325, 329
Causal mask, 205, 206
CF, *see* Collaborative filtering (CF)
Character-level tokenization, 188, 212-214, 310
ChatGPT, 1, 94, 101, 179, 222
classification_report, 268
Classification task, 8, 10-11
Classifier, 93, 98, 99
CNNs, *see* Convolutional neural networks (CNNs)
"Cold-start" problem, 283
Collaborative filtering (CF), 274, 282-283
    embedding vector, 276
    latent factors, 275
    MF (*see* Matrix factorization (MF))
    user-user CF, 275
Collate function, 113, 125, 127, 256
Compute metrics function, 165-167
Computers
    "hear" sound waves, 220, 221
Computer vision (CV), 11, 13, 51, 52
Computing resources, 53, 185
Confusion matrix, 266-272
Convolutional neural networks (CNNs), 53, 90, 224
    activation function, 55
    batch normalization layer, 59, 60, 91
    convolution, 53, 54, 90
    cross-entropy loss/log-loss function, 59, 91
    data augmentation (*see* Data augmentation)
    dimension calculations, 56
    dropout layer, 60, 91
    filters/kernels, 53
    flatten layer, 58, 91
    fully connected/dense layer, 58
    horizontal edge detection, 54
    padding, 56
    pooling layer, 57, 58, 90
    softmax function, 58, 91
    strides, 56
    vertical edge detection, 54
Convolution layer, 52, 55-57, 71, 72
CREMA-D, 271
cross_entropy function, 199
CrossEntropyLoss function, 117, 122, 217
CV, *see* Computer vision (CV)

D

Data augmentation, 60, 91, 254, 255, 271
    accuracy, 86
    CNN model, fourth version, 85, 86
    data statistics, 83, 84
    loading data, 82
    Oxford-IIIT Pet dataset, 84, 87
    sample images, 84, 85
    techniques, 61
    test data, 82
    torchvision.transforms module, 81
    transformations, 81, 82
Data collection, 22
Data exploration, 64-66
DataLoader, 63, 64, 113, 233, 257, 314, 315, 317, 335
Datasets, 186, 187
Data visualization, 64-66
Decibels (dB), 224
DecoderBlock, 183, 203, 204

INDEX

Decoder-only transformer, 177, 216
Decoder-only transformer architecture
    decoder block, 183
    logits, 184
    multi-head attention, 182
    positional embedding, 182, 183
    self-attention mechanism, 179–182
    token embeddings, 183
Decoder-only transformer model
    architecture
        AttentionHead, 203–205
        DecoderBlock, 203, 204
        LanguageModel, 203, 204
        MultiheadedAttention, 203–205
    model config, 202, 203
    testing, 211
    text generation, 212
    training, 211
    word-level tokenization, 212–214
Deep learning, 87, 97
Deep learning models
    hidden state, 179
    LSTM, 179
    RNNs, 178
Digital waveforms, 221
DistilBERT model, 93, 128–133, 141

E

EarlyStopper class, 74, 75
EDA, *see* Exploratory data analysis (EDA)
EmbeddingBag layer, 116, 117, 122
Embedding layer, 97, 115, 118
Embeddings, 126, 133
Embedding vector, 276
Encoder, 99
    architecture, 98
    fine-tuning, 99

    pretrained, 99
    scratch, 99
Encoder-decoder architecture, 105, 133, 298, 299, 317
Evaluate function, 197–199
Exploratory data analysis (EDA), 145, 146
    arbitrary example, 159, 160
    fine-tuning pretrained models, 145
    statistics, 160–162
    .str.len() method, 162
    .str.split() method, 162
    training data, 162

F

Feature engineering, 14, 22
Feedback loop, 99
Feed-forward neural network, 7, 183
Fine-tuned model, 90
Fine-tuning, 87, 92, 175
    Hugging Face models, 138, 139
    workflow, 138, 139, 175
Fine-tuning pretrained models, 141
    DistilBERT, 141
    EDA, 145, 146
    emotions dataset, 141
    evaluation metric, 148, 149
    Hugging Face account, 142
    Hugging Face Dataset
        features, 143
        functions, 144
        indexing, 144
        size, 143
    loading dataset, 142
    model inference, 152, 153
    model testing, 151, 152
    model training
        arguments, 150, 151

INDEX

Fine-tuning pretrained models (*cont.*)
    DataCollator, 150
    dictionaries, 149
    model_name variable, 149
    trainer.train() command, 150
  package installation, 141
  tokenization, 147
Fixed-size limitation, 99
flattened_tensor.shape, 259
Flickr8k dataset, 297, 299, 301
Flickr8k.token.txt, 302
forward() function, 322
Fourier transform, 222

G

generate_caption function, 329
Generate text, 193–197
get_batch() function, 190
__get_item__() method, 313
Google Colab, 185, 227, 241
Google's T5 model, 158
GPT, 193
GPT-4, 186
Gradient descent, 5, 15, 17, 22, 36, 38–40, 77
GroupLens, 274

H

Hugging Face, 136
  Dataset Hub, 137
  fine-tuning, 138, 139
  Model Hub, 137
  parts, 136
  pipeline API, 138
  question answering model, 173, 174, 176
Hugging Face Hub, 128, 137, 154, 155, 173

I, J

Image captioning model, 297
  beam search, 334
  decoder's embedding layer, 335
  EncoderCNN and TransformerDecoder, 324
  hyperparameters, 334
  improve performance and quality, 334, 335
  ResNet-50, 334
Image captioning project, 298, 299
  DataLoader, 314–317
  dataset class, 312, 313
  download dataset, 300–302
  encoder using CNN, 317, 318
  exploratory data analysis (EDA), 305–310
  Flickr8k dataset, 299–301
  Flickr8k.token.txt, 302
  forward method, 319
  generate captions for images, 329–332
  human-written captions, 299
  image transforms, 314
  imports, 299
  __init__ method, 319
  model architecture, 317
  model evaluation
    BLEU score, 331, 332
    model testing, 332–334
    SmoothingFunction, 332
  model training, 327–329
  pretrained ResNet model, 319
  read captions, 303, 304
  ResNet-50 model, 318
  spaCy's tokenization, 310
  split, train-test-validation sets, 302
  tokenization and numericalization, 312

training data, 314
transformer-based decoder, 319
    captions embedding, 321, 322
    cross-attention, 321
    image captioning encoder-decoder model, 324–326
    self-attention, 321
    sources of information, 319
    target mask, 323
    transformer decoder layers, 322, 323
validation and test data, 314
vocabulary, spacy model, 311
Image classification, 52, 90
    components, 61, 62, 91
    ConvNeuralNetwork1 model, 66–68
        convolution layer, 71, 72
        convolution-RELU-maxpool, three units, 72, 73
        pooling layer, 71, 72
    data exploration, 64–66
    data loaders, 63, 64
    dataset, 61, 81, 92
    data visualization, 64–66
    history, 52, 53
    hyperparameters, 70
    imports, 62
    loading data, 62, 63, 82
    model training loop, 68, 69
    num_epochs iterations, 70
    results, 71
    test_loop, 69, 70
    training process, 92
        batch normalization layer, 78
        dropout layer, 79, 80
        early stopping, 74, 76, 77
        optimizer improvements, 80
        rate scheduler, 77, 78
ImageNet, 53, 88, 314
ImageNet Large Scale Visual Recognition Challenge (ILSVRC), 53
Industry-standard NLP library, 310
__init__() function, 313, 319, 325
input_encodings, 114
Iterable-style dataset, 106

K

k-Nearest Neighbor (k-NN) algorithm, 282, 283, 287, 288

L

Language models
    batching, 190, 191
    bigram model, 192–202
    block size, 214
    computing resources, 185
    datasets, 186, 187
    decoder-only transformer model, 202–212
    deep learning models, 178, 179
    diverging loss, 215
    imports, 185
    model capacity, 214
    n-gram model, 178
    out of memory, 215
    probabilities, 177
    subword tokenizer, 215
    text sequence, 178
    tokenization, 188–190
    train for longer, 215
    train-test-validation split, 188
    transformer decoder-only architecture, 179–184
    transformers, 179
    troubleshooting, 215

Latent factors, 275, 276, 296
Layer normalization, 103, 183, 184, 208
Learning rate, 17, 77, 169, 237
Linear layer, 184, 208
Linear regression, 2–5, 13, 14, 20
load_state_dict() function, 266
Long short-term memory (LSTM), 99–100, 178–179, 271

M

Machine learning (ML), 1, 136, 219
    AI, 2
    algorithm, 5
    classification task, 10
    domain, 11
        data preprocessing, 12
        datasets, 12
        labeling, 12
    features, 4, 8
    gradient descent, 5
    hidden layers, 7
    input/training dataset, 4
    linear regression, 2
    mean squared error, 11
    model training, 5
    multiple factors, 4
    parameters, 3
    prediction problem, 2
    rectified linear unit, 6
    ReLU, 6
    setting, 8
    sigmoid and tanh activation, 6, 7
    traditional, 12
    workflow, 10
Map-style dataset, 106
Matrix factorization (MF), 275, 290–295
Mean squared error (MSE), 3, 5, 11
Mel filter banks, 223
Mel scale, 223–224
Mel spectrogram, 219, 221
    for classification, 224
    digital audio waveform, 222
    Fourier transform, 222
    frequencies, 222
    Mel scale, 223
    STFT, 222
ML, *see* Machine learning (ML)
MobileNet, 53
Model architecture, 66, 72, 98, 126, 131–132, 192–193, 203, 234, 258, 317
model.state_dict(), 265
Model training, 5, 15, 98–99, 149–151, 199–201, 211, 261–265, 326–328
MovieLens dataset, 276, 277, 282
MSE, *see* Mean squared error (MSE)
Multi-head attention, 182–184
Multi-modal models, 297
    classification and reasoning, 298
    encoder-decoder architecture, 298, 299, 317
    generation, 298
    information retrieval, 298

N

Natural language inference (NLI), 139, 140, 176
Natural language processing (NLP), 11, 13, 93, 94, 110, 132
    applications, 177
    ChatGPT, 94
    deep learning, 97
    embedding layer, 97
    encoder, 98

encoding, 95
foundational task, 177
preprocessing, 95
problems, 94
tasks, 136, 176
text classification, 97
Neural networks, 7, 15, 16
  gradient descent, 38–40
  parameters, 36
Neuron, 6, 7, 52
N-gram models, 178
NLI, *see* Natural language inference (NLI)
NLP, *see* Natural language processing (NLP)
nn.Embedding, 116, 117, 192
nn.EmbeddingBag, 114, 116–118
Numericalization, 96, 111, 312
NumPy array, 32
num_time_frames, 230, 234
Nyquist-Shannon theorem, 221

O

One-hot encoded vectors, 97
Outliers, 13
Overfitting, 19–21

P

Padding, 56, 96, 327
Perplexity, 199, 217
Positional embedding, 103, 182, 183, 216, 321
Preprocessing, 110
  input encoding, 110
  numericalization, 111
  tokenization, 110
Pretrained models, 87–90, 136–139, 175

Pretrained Word2Vec model, 123, 124
PyTorch, 61, 67, 77, 87, 96, 106, 116, 295, 322, 323
  tensors (*see* Tensors)
  torch.nn module, 91
  torch.optim module, 77
  torchvision.datasets module, 61

Q

Quantization, 221
Quitar string, 219, 220

R

rand() function, 32
RavdessSpeechDataset class, 245, 246
raw_sample, 231
Raw waveform, 221
Recall-Oriented Understudy for Gisting Evaluation (ROUGE), 164–167
Recommender systems, 273
  CF (*see* Collaborative filtering (CF))
  content-based filtering, 274
  platforms, 273
  web-based consumer companies, 273
Recommender systems project, 276
  data analysis and visualization, 279, 281
  imports, 276, 277
  ML class, 290, 291
    evaluation loop, 294
    hyperparameters, 292
    training loop, 292
    user and item embeddings, 295
  MovieLens dataset, 277–279, 282
  user–user CF, 282
    evaluation, 289

INDEX

Recommender systems project (*cont.*)
    ratings prediction using k-NN, 288
    set item averages, 287
    steps, 282, 283
    user-item matrix, 283–285
    user similarity matrix, 285–287
Recurrent neural networks (RNNs), 99, 178
  architecture, 100
  bidirectional, 100
  drawback, 100
ReduceLROnPlateau scheduler, 237, 262
Regression task, 8, 11
Regularization, 20, 57
ReLU activation, 6, 15, 72, 78, 79, 208, 234
Residual connection, 183, 184, 209, 323
resnet18 model
  accuracy, 90
  architecture, 87
  building blocks, 87
  hyperparameters, 88
  ImageNet dataset, 88
  requires_grad, 90
  sizes of layers, 87, 88
  torchvision.models module, 87
  training/validation/testing loops, 89
ResNet-50 model, 90, 297, 313, 318, 319, 334
ResNets, 53
RNNs, *see* Recurrent neural networks (RNNs)
ROUGE, *see* Recall-Oriented Understudy for Gisting Evaluation (ROUGE)
Ryerson Audio-Visual Database of Emotional Speech and Song (RAVDESS) dataset, 219, 242, 244, 271

S

Scaled dot-product attention (SDPA), 181, 182
Self-attention, 101–103, 179–182, 216
Self-attention mechanism
  Key (K), 180
  Query (Q), 179
  SDPA, 181, 182
  tokens, 180
  Value (V), 180
  vectors, 180
Self-feeding loop, 193
sentence_bleu function, 334
Sequence-to-sequence models
  decoder, 104
  encoder, 104
SER, *see* Speech Emotion Recognition (SER)
Short-time Fourier transform (STFT), 222
should_stop function, 75–77
SmoothingFunction, 332
Softmax function, 58, 72, 91, 102, 115, 181, 206
Sound wave, 219–222
spaCy model, 310, 311
SpecAugment, 254–256
Spectrogram, 222–224, 228
Speech Command dataset, 224–226
Speech command recognition, 219, 224, 272
  AmplitudeToDB transform, 228
  Auxiliary Words, 225
  _background_noise_, 226
  commands, 225
  data preprocessing, 231–234
  dimensions, Mel spectrogram transformation, 230

EDA, 227
hyperparameters, data processing and training, 227
imports, 225
label mapping, 231
Mel spectrogram, 224, 229, 230
model architecture, 234
model training and evaluation
    evaluate function, 238
    model setup, 237
    testing, 241
    training loop, 239
Resample function, 228
wanted_words, 231
Speech Emotion Recognition (SER), 219, 241–243
    imports, 242
    RAVDESS dataset, 242, 244
        audio processing parameters, 245, 246
        confusion matrix, 268–272
        custom collate functions, 255
        custom dataset class, 245
        data augmentation, 254, 255
        define transforms, 250
        exploratory data analysis, 250–252
        handle data imbalance, 257
        __init__(), __len__(), and __getitem__(), 245
        label-to-index mapping, 246
        model testing, 266
        model training, 261, 265
        precision, recall and F1-score, 267, 268
        speaker-independent splits, 253
        SpecAugment, 254, 255
        Train-Test-Validation Split, 253, 254
        speaker-independent K-fold cross-validation methodology, 271
        transformer encoder, 271
STFT, see Short-time Fourier transform (STFT)
Storytelling language model, 214, 216
Subword tokenizer, 215

T

Tabular data, 12
Tapestry, 274
Tensor manipulations
    Boolean mask, 42, 43
    exercises, 40, 42
    functions, 43
    image datasets, 45
    integers, 42
    1D sensors, 46–48
    torch.argmax function, 46
    torch.clamp function, 44
    torch.squeeze(), 47
    2D sensors, 48
    2D tensors, 43, 44
Tensors, 48
    attributes, 34, 48
    backward() function, 36, 38
    c.grad/e.grad, 37
    computational graph, 38
    creation, 32, 33
    definition, 23
    functions, 33
    .grad attribute, 38
    grad_fn, 36
    IndexError, 25
    initialization, 48
        from another PyTorch tensor, 31
        constant values, 31

INDEX

Tensors (*cont.*)
    is_leaf attribute, 37
    manipulations (*see* Tensor manipulations)
    NumPy array, 32
    one-dimensional tensor (1D tensor), 24
    operations, 34, 35, 49
    physical memory, 30, 31, 48
    rand() function, 32
    random sampling, 33
    requires_grad, 36, 38
    slicing, 27, 28, 30
    sub-tensor, 48
    three-dimensional tensor (3D tensor), 26, 27
    two-dimensional tensor (2D tensor), 24, 25
    types, 37
TESS, 271
Test loop, 69–70, 120
Text classification, 105
    dataset, 105, 106
    fine-tuning pretrained models (*see* Fine-tuning pretrained models)
    imports, 106
    zero-shot classification, 139, 140, 176
Text generation, 94, 158, 202, 203, 212
Token embeddings, 180, 183
Tokenization, 95, 96, 110, 147–148, 162–164, 188–190, 212–214, 310
token_list, 127
torch.device() function, 227
Trainer class, 175
Training data, 9, 13, 20, 81, 99, 127, 162
Training loop, 118, 121
train_loop function, 68, 70, 78

Train-test-validation split, 8–10, 188
Transfer learning, 87–89
Transformer, 101, 128
    architecture, 179
    decoder-only transformer architecture, 179–184
Transformer-based encoders, 103
    layer normalization, 103
    positional embedding, 103
Transformer models, 156, 175, 202, 203, 211

U

Underfitting, 21, 22
Unigram precision, 331, 334
unknown category, 225
User-item matrix, 283–285
User similarity matrix, 285–287
User–user CF, 275, 282–283
    See also Recommender systems project

V

Validation data, 9, 13, 74, 121
Validation loops, 76, 119–121

W, X, Y

WeightedRandomSampler, 257, 272
Word2Vec model, 123–127
Word-level tokenization, 177, 212–214, 217

Z

Zero-shot classification, 136, 139–141, 176

GPSR Compliance

The European Union's (EU) General Product Safety Regulation (GPSR) is a set of rules that requires consumer products to be safe and our obligations to ensure this.

If you have any concerns about our products, you can contact us on

ProductSafety@springernature.com

In case Publisher is established outside the EU, the EU authorized representative is:

Springer Nature Customer Service Center GmbH
Europaplatz 3
69115 Heidelberg, Germany